HUMAN FACTORS IN AUDITORY WARNINGS

HUMAN FACTORS IN AUDITORY WARNINGS

Edited by
NEVILLE A. STANTON
JUDY EDWORTHY

Ashgate

Aldershot • Brookfield USA • Singapore • Sydney

Published by
Ashgate Publishing Limited
Gower House
Croft Road
Aldershot
Hants GU11 3HR
England

Ashgate Publishing Company
Old Post Road
Brookfield
Vermont 05036
USA

British Library Cataloguing in Publication Data
Human factors in auditory warnings
   1. Signals and signalling 2. Electric alarms 3. Electronic
   alarm systems 4. Electric alarms – Evaluation 5. Electronic
   alarm systems – Evaluation
   I. Stanton, Neville, A. II. Edworthy, Judy
   621.3'8928

Library of Congress Cataloging-in-Publication Data
Human factors in auditory warnings / edited by Neville A. Stanton, Judy
   Edworthy, 1957–
          p.   cm.
          Includes index.
          ISBN 0–291–39849–9 (hb)
          1. Electric alarms—Psychological aspects. 2. Human engineering.
   3. Acoustical engineering. I. Stanton, Neville A., 1960–
   II. Edworthy, Judy.
   TK7241.H85   1998
   621.389'28—dc21   98–3432
                                                               CIP

ISBN 0–291–39849–9

Printed and bound in Great Britain by MPG Books Ltd, Bodmin, Cornwall

Contents

List of Figures

List of Tables

List of Contributors

Dr Chris Baber
Industrial Ergonomics Group
School of Manufacturing and
Mechanical Engineering
University of Birmingham
Edgbaston
Birmingham
B15 2TT
UK

Dr Jim Ballas
Naval Research Laboratory
Code 5513
Washington
DC 20375–5337
USA

Debbie S. Bartolome-Rull
NASA Langley Research Center
Hampton
VA 23681
USA

Dr James Bliss
Psychology Department
University of Alabama at
Huntsville
Huntsville
Alabama 35899
USA

Daniel W. Burdette
NASA Langley Research Center
Hampton
VA 23681
USA

Jennifer L. Burt
NASA Langley Research Center
Hampton
VA 23681
USA

Professor John G. Casali
Auditory Systems Laboratory
Department of Industrial and
Systems Engineering
Virginia Polytechnic Institute
and State University
Blacksburg
VA 24061–0118
USA

J. Raymond Comstock
NASA Langley Research Center
Hampton
VA 23681
USA

Alison F. Cresswell Starr
Smiths Industries Aerospace
Bishops Cleeve
Cheltenham
GL52 4SF
UK

A.J. Datta
MRC Applied Psychology Unit
15 Chaucer Road
Cambridge
CB2 2EF
UK

Dr Judy Edworthy
Department of Psychology
University of Plymouth
Drakes Circus
Plymouth
PL4 8AA
UK

Dr Ellen Haas
Auditory Perception Team
US Army Research Laboratory
Aberdeen Proving Ground
Maryland
USA

Dr Elizabeth Hellier
Department of Psychology
City University
London
EC1V 0HB
UK

Dr Christina Meredith
Division of Psychology
Faculty of Applied Social
Science
Buckinghamshire College
High Wycombe
Buckinghamshire
UK

Dr Jan Noyes
Department of Psychology
University of Bristol
8 Woodland Road
Bristol
BS8 1TN
UK

Dr Roy Patterson
Department of Physiology
University of Cambridge
Downing Street
Cambridge
CB2 3EG
UK

J.A. Rankin
British Airways
Flight Technical Projects
The Compass Centre
PO Box 10
Heathrow Airport
Middlesex
TW6 2JA
UK

Dr Matthias Rauterberg
IPO, Room 1.26
Centre for Research on User-
System Interaction
Den Dolech 2
5612 AZ
Eindhoven
The Netherlands

Dr David Rose
Department of Psychology
University of Plymouth
Drakes Circus
Plymouth
PL4 8AA
UK

Dr Gary Robinson
Auditory Systems Laboratory
Department of Industrial and
Systems Engineering
Virginia Polytechnic Institute
and State University
Blacksburg
VA 24061–0118
USA

Dr Steve J. Selcon
Air Human Factors
Centre for Human Sciences
Defence Evaluation and
Research Agency
Farnborough
Hants
GU14 6SU
UK

Neville A. Stanton Ph.D.
Professor of Design
Department of Design
Brunel University
Runnymede Campus
Cooper's Hill Lane
Egham
Surrey
TW20 0JZ
UK

John R. Welch
Intensive Care Unit
University Hospital Lewisham
Lewisham High Street
London
SE13 6LH
UK

Dr Deborah Withington
Department of Physiology
University of Leeds
Leeds
LS2 9NQ
UK

Preface

Auditory warnings pervade all aspects of our working and domestic lives. In some cases they act as prompts to the performance of a particular action, and in others they call for immediate attention in order to prevent serious mishap. We can plot two orthogonal dimensions – those of intervention immediacy and intervention importance – across which all auditory warnings can be plotted. For example, a warning to a pilot of low fuel would rate high on both intervention immediacy and importance. The appropriateness of a warning along these dimensions will largely determine its success. Many warnings may also change in terms of immediacy and importance dynamically, depending on the context of the situation and the presence of other threats: a warning to a pilot of an approaching missile would relegate a low fuel warning to a much reduced status. Thus warnings also interact with each other as well as with the situations and events to which they refer. In complex systems, patterns of warnings may be indicative of particular problems, and it is the task of the human operator to determine a diagnosis from the presenting symptoms. This can become quite a difficult task in multiple-event scenarios. This and other issues pertinent to the design and evaluation of auditory warning systems are discussed within this book.

The idea for this collection began with a conference hosted by the editors at Chilworth Manor at the University of Southampton on behalf of the UK Ergonomics Society with co-sponsorship from the International Ergonomics Association and the Institution of Electrical Engineers in the UK. The project snowballed, with invitations to leading researchers to contribute to the book, thus giving the text a truly international feel. We would like to thank each of the contributors for making this the first international collection on auditory warnings research and practice. Many thanks also go to John Hindley of Ashgate Publishing for his encouragement and patience. Finally, special thanks go to our respective, long-suffering, spouses and children.

Neville A. Stanton
Judy Edworthy

xvii

PART I
INTRODUCTION

1 Auditory Warnings and Displays: An Overview

NEVILLE A. STANTON, *University of Southampton*, and **JUDY EDWORTHY,** *University of Plymouth*

Introduction

Auditory warnings and displays take many forms. Examples include horns, klaxons, whistles, sirens, bells, buzzers, chimes, gongs, oscillators and a whole host of synthesised sounds, some melodious, some not. There are many characteristics and features associated with the technology of auditory displays, such as intensity, frequency, attention-getting ability, psychological salience and noise penetration ability. Therefore they should be chosen with the background noise environment and the role they are required to fulfil in mind, rather than, as is often the case, arbitrarily assigned to protect a particular device or to warn of certain aspects of that device. Unfortunately 'off-the-shelf' equipment tends not to incorporate these considerations, and even customised equipment is typically produced within tight design bandwidths, there often being very little that the engineer can do outside what the software toolkits and various off-the-shelf devices will allow.

Although there has been much research conducted on the topic of auditory warnings and displays, not all of the problems have been solved by any means. Neither is it the case that the information that already exists has necessarily filtered its way into current design practice. For example, while the design of an auditory display is important, the mere presence of the signal (well or badly designed) is not enough to guarantee appropriate action as there are many psychological and contextual factors which will come into play beyond simple detection of the signal. Auditory signals, especially those which are badly designed, can even became a distraction during times of high workload. The following extracts illustrate some of the problems typically encountered with auditory warnings. The first two are from aviation and the second two are from medical environments.

The 146 has a very comprehensive warning system. We have problems with spurious and unnecessary warnings, especially entering cloud after take off, that is, the worst time. One pilot normally mechanically keeps cancelling the alert. In this instance we had a genuine alert – ice detected – and neither of us reacted correctly until clear of cloud when we had relaxed!

During the approach I was overloaded and received little assistance from the P2 who made a fast approach. As we flared over the runway a strange noise occurred which I could not identify – but did notice that the gear was still up and a down selection stopped the noise just as we came to the hover.

When the alarm kept going off then we kept shutting [the device] off [and on] and when the alarm would go off [again] we'd shut it off ... so I just reset [the device control] to a higher temperature. So I kinda fooled [the alarm].

I know exactly what [the alarm] is ... it's because the patient has been, hasn't taken enough breaths or ... I'm not sure exactly why.

Some of the problems typically associated with auditory displays, particularly warnings, are that there are often too many of them, these are typically too loud and sound too frequently, and they are psychologically inappropiate for a range of reasons. In part, these problems can be traced through the history of auditory warning design. For example, the transfer of alarms from an age of simpler technology to the modern work environment without any adaptation may be inappropriate. Patterson (1989) illustrates this by pointing out that warnings and fire bells that may have been appropriate for steam locomotives with open cabs are inappropriate for today's trains, even though they may still be present in the cabs of these modern trains. The sheer number of auditory warnings often used in a single environment can also be traced historically, as they tend to be added piecemeal without much consideration being given to the system as a whole. Thus a system which starts out with only a few warnings may, over a period of a few years, develop into one where literally hundreds of warnings may be in use. Not only is this a problem for the operator, who cannot hope, even with the best training, to be able to distinguish between these warnings; it will also increase the potential for masking, whereby one warning may not be audible because another is sounding at the same time, and is masking it acoustically as a consequence.

The problems seem largely to originate from a 'better safe than sorry' philosophy of alarm design (Patterson, 1985). Considered in isolation, a warning may be sufficiently salient to call attention to a problem so that the probability of a miss is low. This is important in practical terms because the consequence of a miss may be quite drastic in terms of plant, product or personnel (Stokes *et al.*, 1990). False alarms, which such a philosophy is almost bound to create, not only create mistrust of the warning system (Bliss *et al.*, 1995), but can also create a buzzing, flashing confusion of sound and lights when an incident actually occurs. This would hardly seem conducive to calm fault management. Patterson (1990) illustrates this last point by citing part of a confidential incident report. This is reproduced below.

> I was flying in a Jetstream at night when my peaceful reverie was shattered by the stall audio warning, the stick shaker and several warning lights. The effect was exactly what was NOT intended; I was frightened numb for several seconds and drawn off instruments trying to work out how to cancel the audio/visual assault rather than taking what should have been instinctive actions.... The combined assault is so loud and bright that it is impossible to talk to the other crew member, and action is invariably taken to cancel the cacophony before getting to the actual problem.

Given that the fault management task can be very time critical in an aircraft, it is clear that, from a human factors point of view, the pilot was not supported as well as he or she might be by the alarm instrumentation.

The Use of Auditory Warnings and Displays

One unquestionable benefit of auditory warning and display systems is that they present a means of unburdening the visual channel. However overuse can lead to auditory clutter and the signals can, if not properly designed, be unpleasant and intrusive. For example, during the Three Mile Island power plant crisis more than 60 different auditory warning systems were activated. Such systems are not ergonomic and tend to restrict, rather than enhance, performance.

Auditory warnings and displays can command the attention of most people within the vicinity by breaking through background noise. This is perhaps most frequently encountered by the public in situations where the emergency services are in operation. Under such conditions the sirens of ambulances, police cars and fire engines attract attention by cutting through the noise of the traffic, but do so through sheer brute force. In such situations, design is focused on maximising the 'attention-getting' aspect of the auditory medium (McClelland, 1980). Often the same philosophy of flooding the environment with sound to attract attention is transferred to situations where attention needs to be drawn, but it is done in such a manner that it is more likely to startle than cue the appropriate action (Patterson, 1990). This is not helped by a design ethos which until a few years ago assumed that the purpose of an auditory warning was to startle; nowadays it is recognised that to do so, especially when an operator is working hard and is under stress, is not ergonomically justified and that it is often better to design warnings which attract attention in a less disruptive way.

Table 1.1 illustrates some of the potential benefits of auditory displays over visual displays. In general, auditory displays require little directional search, responses tend to be faster than to visual displays, urgency mapping and prioritisation are relatively easy to incorporate, they are not affected by visual noise, they are flexible in terms of user mobility and they tend to be most suited to signalling time-dependent information. Auditory displays do appear to offer some substantial benefits over visual displays; for example Stokes *et al.* (1990) suggest that auditory alarms alert operators quickly, irrespective of head or eye position, and appear to do so more quickly than do visual displays. On the other hand, information is often hard to convey through non-verbal auditory displays in particular, and people's sheer processing power is restricted for auditory information as compared to visual information. But because we

Table 1.1   Relative advantages of the alarm display media

	Auditory display	Visual display
Reception	Requires no directional search	Requires attention and selection
Speed	Fastest	Slowest
Order	Difficult to retain	Easy to retain
Urgency	Easy to incorporate	Difficult to incorporate
Noise	Not affected by visual noise	Not affected by auditory noise
Accepted symbolism	Melodious, linguistic	Pictorial, linguistic
Mobility	Most flexible	Some flexibility
Suitability	Time-dependent information	Space-dependent information

cannot 'shut our ears' in the same way that we can our eyes, our hearing tends to act as a natural warning sense. It is no surprise, then, that auditory displays and alarms are commonplace.

Another contrast between the auditory and the visual medium comes from Deatherage (1972, cited in Sanders and McCormick, 1987). This is reproduced in Table 1.2 below. Sanders and McCormick suggest that the auditory medium is preferable to the visual medium when

- the origin of the signal is itself a sound;
- the message is simple;
- the message will not be referred to later;
- the message refers to events in time;
- the message calls for immediate action;
- continuously changing information of the same type is presented;
- the visual system is overburdened;

Table 1.2   When to use the auditory or visual form of presentation

Use auditory presentation if:	Use visual presentation if:
The message is simple	The message is complex
The message is short	The message is long
The message will not be referred to later	The message will be referred to later
The message deals with events in time	The message deals with locations in space
The message calls for immediate action	The message does not call for immediate action
The visual system is overburdened	The auditory system is overburdened
The receiving location is too bright or dark adaptation integrity is necessary	The receiving location is too noisy
The person's job requires moving about continually	The person's job allows them to remain in one position

Source: Sanders and McCormick (1987) citing Deatherage (1972).

- speech channels are fully employed;
- illumination limits vision;
- the receiver moves from one place to another.

Such circumstances are likely to dictate that auditory warnings and/or displays are advisable. It may be inferred that, if the opposite of these conditions pertains, then the use of auditory alarm displays would not be advisable. Thus visual displays might be more appropriate if

- the origin of the signal is not itself a sound;
- the message is complex;
- the message will be referred to later;
- the message refers to spatial locations;
- the message calls for delayed reaction;
- information of different types is presented;
- the auditory system is overburdened.

To a large extent, these recommendations will hold for future applications because they are based on the contrasting and complementary nature of the visual and the auditory senses. However the use of auditory displays in the past has been restricted by the technology available and the dearth of knowledge of psychoacoustic and psychological processes associated with the use of such displays. As there have been considerable advances in both areas (such as the use of icons and environmental sounds, three-dimensional auditory displays, urgency mapping and the interpretation of false alarm data), the areas of application and use of auditory displays and warnings are likely to increase and shift more towards being capable of replacing the visual sense in applications where a visual would have traditionally been the more obvious mode. One example would be the use of three-dimensional auditory displays, which are ideally suited to signifying information in particular spatial locations even though guidelines have traditionally suggested that auditory displays are not efficiently used in this kind of application.

The Design of Auditory Displays

Auditory alarm and warning displays are commonplace, and may be divided loosely into four classes of areas of application. These are personal devices, transport, military and central control rooms. Personal devices include alarms clocks, anti-rape alarms and burglar alarms. These devices are personal in that they are intended for use by one individual and are not part of a wider system. Transport applications include cars, buses and civil aircraft. These are different to the personal devices in that they typically possess more than one alarm, and may also be multi-person systems. Military applications include missiles, armoured fighting vehicles and fighter aircraft. These are different from the civil applications in that there are different demands placed upon the operator owing to the nature of the task being performed. They also require highly specialised training, and typically an individual could not transfer simply from the civil to the military task. Warning systems in military applications tend also to be more highly developed, because they have to deal with threats from outside the operation of the machine itself as well as internal failures. Central control room applications include coronary care units, manufacturing and power generation plants. These are again different to the other areas of application as they typically involve team supervision of complex processes that tend to be monitored from rooms elsewhere in the building. Here there is a particular problem of inferring causality from raw data taken from a variety of sources.

Such is the range of applications of auditory warnings and displays that it would be impossible to propose simply one design methodology and implementation policy. However, while to some extent the particular application and its context might influence the actual design and use of the auditory signals, there are some basic principles which need to be adhered to across the whole sphere of auditory warning and display design. For example, the very first requirement of an auditory display or warning is that it be heard, and the second is that its meaning be understood. Understanding will be achieved partly through training, partly through design and partly through the number of signals in a system (as even the best designed auditory warning or display set is likely to be hard to understand if dozens, or even hundreds, of signals are used). Ensuring that this is the case is not a trivial problem and requires the application of a substantial body of knowledge to the process of design and implementation. However, even if these two crucial criteria are met, there is much more to auditory warning design and implementation than simply ensuring that sounds are heard and understood, as we shall see later in this chapter. The whole sequence of actions from detection to action needs to be mapped and understood (Stanton, 1994). Typically work in applied areas of warning and display design has focused on the early stages of detection and interpretation. In the next section we deal with these issues and review some of the approaches that have been taken to the problems of detection and meaning.

Warning Detection

For practical purposes the detectability of a sound is not influenced by people's absolute threshold, but by the masked threshold of the environment in which the operator is working. That is, account needs to be taken of the ambient background noise in which the sounds will typically be heard. In some areas of application this noise level will be fairly fixed, and in others it will fluctuate. In fluctuating noise environments it will inevitably be the case that the sounds will sometimes be louder than they need to be, so decisions will have to be made at some stage in the implementation of an auditory warning and/or display system concerning the extent to which potentially missed signals might compromise safety.

One of the features of auditory perception is that a person's ability to hear a sound is dependent largely on other sounds around the same frequency value; sounds that are close in frequency to a signal are much more likely to mask that signal than potential maskers which are farther away in frequency. Thus it is important to take account of the intensity pattern across the whole of the noise spectrum before predictions can be made about a signal's detectability. The auditory filter provides a comprehensive and well-researched model which serves a crucial role in determining signal detectability, and serves as the theoretical backdrop to auditory warning and display detectability. There are at least two comprehensive and detailed approaches to alarm detectability (Patterson, 1982; Laroche *et al.*, 1991) and both use models of the auditory filter as a basis for their predictions and recommendations (Patterson, 1974, 1976, Patterson and Nimmo-Smith, 1980, in the case of Patterson (1982), and Zwicker & Scharf, 1965, in the case of Laroche *et al.* (1991)). These approaches are described in detail elsewhere (Edworthy and Adams, 1996, ch. 5). In brief, both models and approaches can establish whether or not, for a given noise environment, a sound will be heard in relation both to the noise environment itself and to the other sounds with which it is likely to be heard. Patterson (1982) states that a sound will be heard reliably if it is 15dB above masked threshold, which can be predicted from a combination of noise measurements and superimposition of the auditory filter across the whole frequency range, and by 25dB above this threshold it will be hard to miss, so there is little point in exceeding this value (to do so will render the sounds unnecessarily loud, with all the consequences that are likely to follow). Laroche *et al.*'s model, which can be obtained as a computerised expert system, demonstrates potential masking by superimposing the excitation pattern which would be produced by one sound over another; if one pattern completely covers another, the sound which is covered will not be audible. Laroche *et al.*'s model also includes refinements such as hearing loss and the use of hearing protectors, and so is applicable in a wide range of environments.

Such is the complexity of the detectability of sounds in noise that it is not enough simply to approximate the level of auditory warnings and displays by listening to those sounds in the application in which they are to be used. This will often lead to sounds being too loud in a specific frequency range, which in turn implies the risk that they will become aversive to the extent that they might either be permanently turned off or that alarm cancellation becomes a primary task when a crisis occurs.

Meaning of Sounds

Aside from the impact of training, which will almost inevitably lead to better understanding of auditory warnings and displays, two factors which will influence people's ability at least to understand the meaning of sounds are, first, the design of those sounds themselves and, second, the sheer number of sounds used in a particular application. The first of these problems is the more interesting theoretically, whilst the second is one to which there are fairly straightforward answers, but one which in practice often becomes a problem, as the number of sounds used within a system often escalates. For example, it is technically easy to equip a single piece of machinery with several different signals; one or two might indicate the actual problem being monitored, another set might indicate specific malfunctions related to the machinery rather than the problem itself, and yet others might inform the operator that the machinery has just been switched on or is performing a self-test. While the technology needed to create all these sounds is straightforward nowadays, the psychological impact of all these sounds is a much more complex issue and has yet to be fully understood.

Types of Sound

Nowadays it is possible to synthesise almost any sort of sound, from an abstract, 'traditional' type of auditory warning to a sound image or an exact copy of an environmental sound. There is no agreed consensus over which types of sound are best suited for use as auditory warnings and/or displays, and a range of types of sounds will be discussed in this book. It may well be the case that there is some kind of interaction between the type of sound used and the number which can successfully be implemented in a sound set. For example, as will be shown later in this chapter, people's ability to learn and remember a set of abstract alarms is severely limited. On the other hand, people are capable of understanding literally hundreds of sounds in their normal environment, so sounds which are more naturalistic may lead to better retention rates. Whether such sounds can function as warning sounds in particular is an interesting topic for research. While there is a lengthy tradition of using abstract sounds as warnings for specific, highly urgent situations, the application of more naturalistic sounds in auditory displays is an area in which there is much potential for research and development.

Almost all recent proposals for the design of auditory warning and display systems reject the old idea of using traditional sounds such as bells, horns, klaxons and so on in warning and display systems. One of the earliest counterproposals was put forward by Patterson (1982). He proposed that auditory warnings should gain attention, but not startle (as is traditionally the case) and they should allow communication to take place, which traditional warnings do not generally allow. He also took account of the fact that the most confusing feature of traditional alarms appeared to be their repetition rate, and therefore proposed that each warning in a set should have a distinct temporal pattern. Specifically he proposed that an auditory warning might be constructed from bursts of sound that can be repeated at varying intervals bearing a relation to the urgency of the required action and the level of background noise. A burst is a set of pulses which give a syncopated rhythm – a melody that can be used to identify both the nature and the urgency of the warning. The basic building blocks of the bursts are tones of 100–300 ms in length, whose spectral and temporal characteristics can be matched to the noise environment. They also have a shaped onset envelope in order to reduce startle responses. This definition is taken from Patterson (1990) and the construction of the three elements, warning, burst and pulse, are illustrated below. This type of construction is proposed as one means of developing an 'ergonomic' warning sound, that is one that is informative but does not unduly startle the operator.

In Figure 1.1 the complete warning is represented in the bottom row, where the warning contains bursts of sound numbered from I to VI. The spectral and temporal characteristics of the pulse (shown in the top row) and the bursts (shown in the middle row) give the warning its distinctive character. Patterson (1990) indicates that the perceived urgency communicated by the warnings may be altered by adjusting the pitch, intensity and speed of the burst. Indeed one of the benefits of such a design is that it allows the sounds used to be manipulated in terms of their urgency, which is not possible with more traditional designs. The topic of perceived urgency has received a lot of attention in the literature (for example, Edworthy *et al.*, 1991; Hellier *et al.*, 1993) and is considered later in this book. One of its

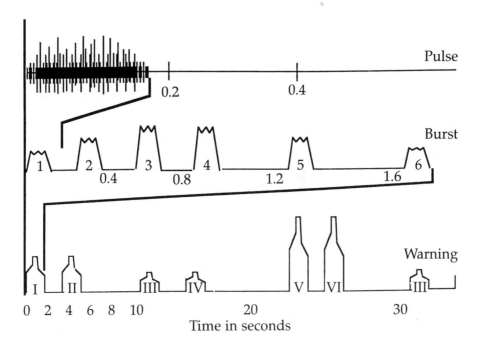

Figure 1.1   Construction of an auditory warning

Source: Patterson (1982, 1985, 1989, 1990).

merits is that, along with informing (if the operator has learnt the meaning of the sound), the warning can convey the degree to which the situation or condition being signalled is an urgent, or a non-urgent, one. The benefits of urgency mapping such as this are still in the process of being researched and are also discussed in this volume.

There have been other, potentially more radical, proposals since Patterson's original ergonomic warning design guidelines, many of which focus on the idea that sounds which are more representational, or environmental, might serve to fulfil a useful role in auditory warning and display implementation. Some research from the field of auditory icons, or 'earcons' as they are more popularly called, may provide some insight into this possibility.

Blattner *et al.* (1989), for example, suggest that there may be some principles common to visual symbols (icons) and auditory messages (earcons) which can be used in transmitting information to system users. Earcons have been divided into two classes, in the same way as visual icons, namely 'representational' and 'abstract'. Representational earcons are perhaps the easiest to design in that they will actually sound like the thing they represent, and in this sense they are literally the sounds of the objects themselves. A proviso here is of course that the representational object does indeed have an associated sound, and that the sound will be recognised and easily interpreted by the human operator. The representational category can be further divided. Gaver (1986) investigated 'symbolic' (for example, applause for approval), 'nomic' (for example, the sound of a closing metal cabinet for the closing of a file) and 'metaphoric' (for example, falling pitch for a falling object) mappings for representational earcons. Gaver suggested that representations need not be realistic, but they should capture the essential features of the thing they represent if they are to be successful.

By contrast, abstract earcons require the development of a distinctive audio pattern which appears to be similar to the Patterson model of the construction of auditory warnings described earlier (Patterson, 1982). Both the symbolic and the representational auditory display has the potential to communicate the nature of the fault to the operator, and not just that there is a fault, which is the most that could have been hoped for in more traditional types of display and warnings systems. This potential, if exploited in an intelligent manner, could hold promise for fault management. It could mean reduced time to respond, as the operator does not have to first find and consult visual displays to discover the nature of the problem, and an increase in the amount of message that can be transmitted by the non-speech auditory warning display. Much research still needs to be done but some of the directions in which this research is moving are discussed in this book.

Perhaps the one advantage of traditional types of warnings such as bells, buzzers, horns and so on is that, for many people, associations have been learned between those sounds and the situations and events thay they signal simply through stimulus–response learning. This can be capitalised upon in more recent and ergonomic approaches to auditory warning and display design because sounds can be resynthesised to ensure that they are more ergonomic (by taking out unnecessarily loud harmonics and by shaping onset envelopes, for example) but still retain the essence of the original sound. A study by Lazarus and Höge (1986) suggests that traditional warnings do have specific learned associations for operators. The results of a series of ratings studies show, for example, that the use of a siren fits in well with dangerous situations and horns fit well with mechanical problems. This can be capitalised upon at a psychological level, as resynthesis of such sounds could take out the bad, leaving only the good.

Thus there are many interesting psychological and ergonomics issues worthy of research investigation in this area, and many will be touched upon later in this book. There are many issues which are still quite open-ended, including those as to which sorts of sounds are best suited for particular applications, how certain types of sounds will be interpreted and acted upon, the degree to which the context in which a sound is heard will affect both interpretation and action, and how sets of warnings and displays should be integrated when developing a complete system.

Number of Sounds

There is a considerable amount of evidence to show that the number of abstract sounds that can be retained and recognised by an individual is severely limited. This may not turn out to be the case for representational sounds, so the following section considers abstract sounds only, as they are still the more frequently used auditory warning and display sounds. In a few years' time it may be that number is no longer an issue, at least from a learning and recognition perspective; it will still be an issue in terms of masking, because it will inevitably be the case that the more sounds there are in a system, the more likely they are to signal at the same time, and hence one sound is more likely to mask another. But with the inevitably slow filtering down of research evidence into practice, the problem of number is likely to be pervasive for some years to come.

Patterson (1982) investigated pilots' ability to learn, discriminate and recall the meaning of 10 auditory warnings taken from a selection of 54 warnings on seven civil aircraft. The set of warnings, their meaning and source are shown in Table 1.3. In learning trials presenting the 10 warnings, Patterson demonstrated that four warnings were acquired quite easily, that it took longer to acquire three more warnings, and substantially longer to acquire the last three warnings. Immediate testing showed that recall performance was near perfect, but after a week's delay only seven of the warnings could be recalled. A remedial training session brought this figure up to nine warnings correctly identified. One of the main problems in identification was attributed to the alarm repetition rates. Alarms could be quite different spectrally (that is, they could sound like different 'instruments') and yet if they repeated at approximately the same rates, they would be readily confused. Patterson showed quite decisively that the temporal similarity between warnings strongly influenced the degree to which two warnings would be confused; hence his recommendation that warnings within a set possess distinct temporal patterns. The study by Patterson also suggests that between four and seven warnings can be acquired reasonably quickly; thereafter performance slows down dramatically. Up to seven warnings can be retained, even after one week of absence, and this figure could go up to nine if the warnings were presented regularly.

Table 1.3 Ten warnings taken from civil aircraft

Warning	Aircraft	Verbal description
Fire	BAC 1–11	Ringing bell
Takeoff	BAC 1–11	Intermittent horn
Overspeed	BAC 1–11	Clacker
Undercarriage	L1011	Hollow resonator (horn)
Altitude	L1011	Buzzy, hollow, musical note
Disconnected autopilot	747	Siren being started repeatedly
Selective call	747	Rapidly alternating tones about a minor 7th apart
Glide scope	DC10	Chicken clucking
Passenger	747	Pulses of a shrill bell-whistle
Cabin pressure	L1011	Train of 'bonks' over a swishing background

Source: Patterson (1982).

Usually, although not always, the operator is expected to discriminate between auditory warning displays on an absolute, rather than a relative, basis. This typically is harder to do than discriminating on a relative basis. It is possible to enumerate the number of levels of each of the more important acoustic parameters between which a listener can typically discriminate. This is shown in Table 1.4, which suggests, for example, that people's ability to discriminate between pure tones is from four to seven stimuli on an absolute basis (although more recent auditory warnings research (Meredith and Edworthy, 1994) suggests that the number may be even smaller than that). On a relative basis, discrimination is much finer. The fact that absolute identification might be required should therefore also be borne in mind when designing an auditory warning or display system. Sanders and McCormick (1987), for example, cite Weiss and Kershner (1984) in suggesting that, although 12 auditory alarms may be discriminable on a relative basis, if absolute identification is required then this number is likely to be halved.

Table 1.4   Levels of auditory dimensions identifiable on an absolute basis

Dimension	Levels
Intensity (pure tones)	4–5
Frequency	4–7
Duration	2–3
Intensity and frequency	9

Source: Sanders and McCormick (1987), citing Deutherage (1972) and Van Cott and Warwick (1972).

Design Principles

Sanders and McCormick (1987) provide some general principles for design of an auditory display. These guidelines are presented in three parts: general principles, principles of presentation and principles of installation. These are summarised below.
General principles:

- the signals' characteristics should exploit learned or natural relationships;
- complex information should be conveyed in two parts: an attention-getting part and an information-conveying part;
- signals should be made discernible from any other auditory input;
- do not provide more information than is necessary;
- the same signal should designate the same information at all times.

Principles of presentation:

- avoid extremes of auditory dimensions;
- establish intensity relative to ambient noise level;
- use interrupted or variable signals;
- do not overload the auditory channel.

Principles of installation:

- test the signal to be used;
- avoid conflict with previously used signals;
- facilitate the changeover from the previous display.

Patterson's 1982 guidelines focus on the acoustic and psychoacoustic properties of a warning, and have the following features:

- the lower limit for the warning sound should be 15 dB above the background noise;
- the upper limit for the warning sound should be 25 dB above the background noise;
- the pulses of sound used to build a warning sound should have onsets and offsets that are 20–30ms in duration to avoid startle reactions;
- the sound pulses should be 100–200ms in duration;
- the appropriate frequency region for the spectral components (at least for flight deck warnings) is 0.5–5.0kHz.;
- the warning sound should contain more than four components and the components should be harmonically related so that they fuse into a concise sound;
- the fundamental should be in the range of 150–1000Hz, and at least four of the prominent components should fall into the range of 1.0–4.0kHz;
- for immediate action warnings the sounds might contain a few quasi-harmonic components and/or a brief frequency glide to increase the perceived urgency of the sounds;
- the warning should be composed of five or more pulses in a distinctive temporal pattern to minimise the probability of confusion among members of the warning set;
- for urgent sounds the inter-pulse interval should be less than 150ms; for non-urgent sounds the interval should be over 300ms;
- manual volume control should be avoided;
- there should be no more that six immediate action warning sounds and up to three 'attensons' (attention-getting sounds).

Since the time of Patterson's original guidelines there has been some relaxing of the design principles, although the acoustic principles are, by and large, retained. Developments in this area are discussed later in the book. Specifically with regard to the acoustics of warning and display design, but on a more general basis, Sanders and McCormick (1987) have proposed the following guidelines:

- use frequencies of between 200 and 5000Hz;
- use frequencies below 1000Hz if the signal must travel any distance;
- use frequencies below 500Hz if there are obstacles;
- use a modulated signal;

- use signals with frequencies different from the background noise;
- make warnings discernible from each other;
- use a separate communication system for warnings.

Beyond Detection and Understanding

Although most of the focus of interest in research into auditory warnings and displays has been on the issues of detectability and understandibility (and there is much left to be resolved), there is more to alarm implementation than simply ensuring that a sound is heard and understood. In this section a framework for looking at the whole realm of alarm- and warning-related behaviour is presented.

Little advancement has been made upon the model of alarm handling proposed by Lees (1974) over two decades ago. This model comprised three stages: detection (detecting the fault), diagnosis (identifying the cause of the fault) and correction (dealing with the fault). This model appears to be very similar to the process model put forward by Rouse (1983) comprising detection (the process of deciding that an event has occurred) diagnosis (the process of identifying the cause of an event) and compensation (the process of sustaining system operation). A similar model, comprising detection (detection of the onset of a plant disturbance), diagnosis (diagnosing the particular disturbance from presented symptoms) and remedial actions (selecting and implementing the appropriate actions to mitigate the disturbance), was proposed by Marshall and Baker (1994) as an idealised three-stage decision model to describe how operators deal with faults. There appears to be little to distinguish these three models, apart from the idiosyncratic labelling of the last stage in all cases! Rouse (1983) offers an expanded version of the process model, comprising three levels: recognition and classification (in which the problem is detected and assigned to a category), planning (whereby the problem solving approach is determined) and execution and monitoring (the actual process of solving the problem). Arguably this is reducible to the original three-stage model, but rather more emphasis has been placed upon the interpretation of the problem. What is not clear from any of the analyses presented with these models is that they accurately reflect processes undertaken by the operator within the alarm-handling task. The work of Rouse (1983) is clearly based upon an extensive literature review, but the other models appear to be based upon casual observation.

Activity in the control room may be divided broadly into two types: routine and critical. Incident-handling activities take only a small part of the operator's time, approximately 10 per cent (Reinartz and Reinartz, 1989; Baber, 1991) and yet they are arguably the most important part of the task. This is particularly true when one considers that the original conception of the operator's task was one of operation-by-exception (Zwaga and Hoonhout, 1994). In order to develop a clearer understanding of the alarm handling, a taxonomy of alarm handling was developed (Stanton and Baber, 1995) on the basis of a literature review, direct observation and questionnaire data. This taxonomy reveals 24 alarm-related behaviours subsumed under seven categories: observe, accept, analyse, investigate, correct, monitor and reset (see Figure 1.2).

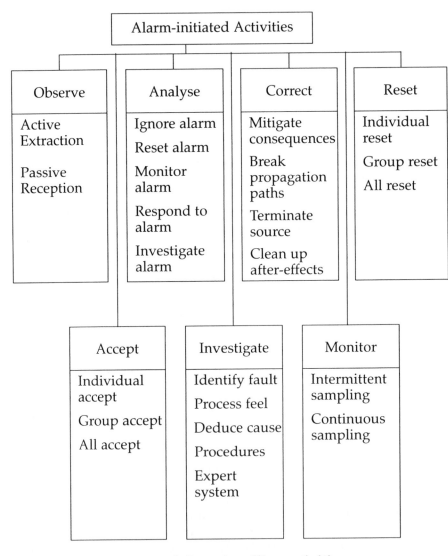

Figure 1.2   Taxonomy of alarm-handling activities

An alarm-handling sequence can be described as consisting of a number of generic activity stages. These have been assembled into an analysis of alarm handling (Stanton, 1994) as shown in Figure 1.3. The analysis distinguishes between routine events involving alarms and critical events involving alarms. Although the two types of event have most activities in common, critical events are distinctive by virtue of an investigative phase. It is proposed that the notion of alarm-initiated activities (AIAs) is used to describe the collective of the stages in alarm event handling. The term 'activities' is used to refer to the ensuing behaviours triggered by the presence of alarms. It is posited that these activities would not have been triggered without the alarm being present, thus they are alarm-initiated activities. The AIAs are linked to other supervisory control activities which can be typified as continuous tasks (such as visual scanning of instruments

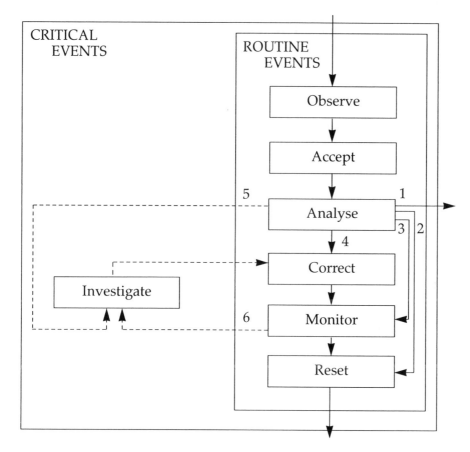

Figure 1.3   Alarm-initiated activities

and fine-tuning of plant variables in response to minor variations in plant) and discrete tasks (such as putting plant into service and taking plant out of service). In such tasks, alarm information may be used instead of, or in conjunction with, other information (such as data on plant variables, some internal reference to plant state, comments from other operators or reports from engineers who are in direct contact with plant). While it may be difficult to distinguish between some activities, whether triggered by alarms or otherwise, examination of alarm-handling activities can be justified in terms of providing useful information regarding the design of alarm systems. This will, of necessity, involve the consideration of activities where alarm information is of primary importance (such as in a critical event) and activities where the alarm information is of secondary and/or supplementary importance to the task (such as in a routine event).

In alarm handling, operators report that they will observe the onset of an alarm, accept it and make a fairly rapid analysis of whether it should be ignored (route 1), reset (route 2), monitored (route 3), dealt with superficially (route 4) or require further investigation (route 5). If it cannot be cleared (by superficial intervention), they may also go into an investigative mode (route 6). In the penultimate mode, the operators will monitor the status of the plant brought about by their corrective actions and ultimately reset the alarm. Routine behaviour has a 'ready-made' response, whereas critical behaviour needs knowledge-based, deductive, reasoning. The taxonomy (observe, accept, analyse, investigate, correct, monitor and reset) is proffered as a working description of alarm-handling behaviour, rather than being intended to represent a fully validated psychological model.

The development of a taxonomy and model of alarm handling provides a focus on the design formats required to support AIAs. The human requirements from the alarm system may be different, and in some cases conflicting, in each of the seven categories of the taxonomy. The difficulty arises from the conflicting nature of the stages in the model, and the true nature of alarms in control rooms; that is, they are not single events occurring independently of each other but are related, context-dependent and part of a larger information system. It is not easy to separate alarm-initiated activity from general control room tasks, but it is useful to do so. While many of the activities may also be present in more general aspects of supervisory control tasks, their consideration is justified by relating them to the alarm system in particular. This leads to the following observations. First, designers of alarm systems need to consider all manner of AIAs. Often there is little consideration of alarm handling beyond the initial 'observe' stage. Second, one could question the need to consider the alarm system as a separate entity from the other displays in the control system.

Human Information Processing and Auditory Displays

Auditory Attention

Wickens (1990) suggests that the short term auditory store retains information longer than the corresponding visual store. Theories of auditory attention suggest that auditory input may be retained in a pre-attentive short-term auditory store for a period of between three and six seconds (Wickens, 1990). This means that the contents of the store may be examined retrospectively, providing that attention is switched back in time. Even if not attended to, this information may be examined at a pre-attentive level. If the information is deemed to be sufficiently salient (for example, a sudden environmental change such as a loud auditory warning), it will be brought to our attention. In the case of auditory warnings and displays specifically, it is important not to produce too loud a sound as, although it will get the listener's attention, it may do so at the expense of their ability to get on with the task in hand. In many instances the listener may be working in an environment where a lot of noise is present, which will also have detrimental effects on performance, as discussed later in this section. Therefore with sound design it is important to balance the attention-getting quality of the sound with the impact that an unnecessarily loud sound will have on performance.

Although it may be possible to pay attention to the different dimensions of auditory signals and selectively to filter out sounds which are different in pitch or other content, storage of two or more sounds can become more problematic if there is over 10dB difference between them, or if they are presented from the same spatial location. Sound also suffers from a phenomenon known as 'visual dominance', whereby there is a general bias towards a preference for visual information if the same information is presented in both visual and auditory modalities. However, Wickens (1990) cites a study by Posner *et al.* (1976) which illustrates the superiority of the auditory mode when presenting information designed to alert, which is what auditory warnings and displays are usually intended to do. This study demonstrated that, whilst an auditory warning led to a quicker response for a subsequent auditory or visual stimulus, a visual warning did not.

Where possible, the auditory system searches for the most parsimonious resolution of an auditory stimulus; our hearing will group sounds according to similarities in frequency, temporal rate, timbre and rhythm. This type of perceptual organisation may decrease the cognitive load and increase the efficiency with which information can be processed (Stokes *et al.*, 1990). A phenomenon known as 'pitch streaming' (Bregman, 1990) is one such organisational phenomenon and shows that listeners will group sounds by pitch proximity even when other information (such as location or timbre) suggests that there is more than one auditory source. If sounds from two sources are heard simultaneously they can sound as if they are one source, increasingly so as the frequency difference between them is reduced. On the other hand, a single auditory source can be made to sound as if it is coming from two different sources if the pitch separation between them is large enough and the individual pulses or tones are played quickly enough. The latter phenomenon has yet to be exploited in auditory warning and display design, although there may be potential here.

The Effects of Noise on Performance

Anecdotal accounts of alarm handling recount that in some instances procedures require that operators leave the audible warnings ringing (which may often be at a level of between 80 and 90dB) while the fault is being attended to. This practice would seem to be contrary to good human factors practice, on four counts. First, once the warning has attracted attention to the fault its purpose may be served and therefore it may no longer be required. Second, the continuous ringing is likely to be the source of some annoyance. Third, the presence of the noise may use up some attentional resources, in terms of Wickens's (1990) theory. Fourth, there is some evidence to suggest that the presence of background noise, at even quite moderate levels (from 75dB), can significantly impair performance (Smith, 1989). Thus the extensive literature on noise effects is relevant to that on auditory warning and display design and implemention, firstly because the displays themselves are often a considerable source of noise, and secondly because many of the work environments in which such displays are typically used are noisy environments and thus it is important to know how work performance might be affected. In general, noise acts as a non-specific stressor and can have auditory, non-auditory, behavioural, communication and social consequences (Edworthy, 1997). While the auditory (for example, noise-induced hearing loss), non-auditory (for example hypertension), communication and social consequences of noise are important, in this section we focus on the effects of noise on performance as it is likely to be the most important practical and research issue in relation to noise and auditory warnings and displays.

As a non-specific stressor, noise will increase our level of arousal, which in turn will affect our ability to perform tasks. In some instances noise may actually improve performance, although generally it will impair it. Much of the research on the effects of noise on task performance is necessarily low on ecological validity; most of the studies reported are conducted in the experimental laboratory using very simple tasks. However these very strict conditions are necessary because noisy environments are often confounded with other variables which may also affect performance, and so laboratory studies allow control over these potentially confounding variables. Although there has been a tendency for the research to concentrate on the effects of high-level continuous noise on relatively simple manual tasks, some research has also considered the effects of intermittent noise on cognitive tasks (Smith, 1989, cites Broadbent, 1979).

Intermittent noise is perhaps more representative of the types of environment in which auditory signals might be presented within human supervisory control tasks. The studies conducted consistently report performance being disrupted if the noise is present when there is an intake of information or where a response is to be executed (Smith, 1989). Perhaps surprisingly, though, the presentation of noise does not appear to disrupt the processing of information (Smith, 1989, cites Woodhead, 1964). The effects on performance produced by intermittent noise seem to be 'local' to the time period following the onset and offset of the noise. The extent to which performance is disrupted also seems to depend upon the degree of change in ambient noise level: the greater the change, the worse performance becomes (Smith, 1989 cites Teichner *et al.*, 1963). Short duration noise exposure of moderate intensity may produce biases in the allocation of processing resources. This has a knock-on effect of making the operator rather inflexible and less adaptable to change, which ultimately reduces their efficiency in tracking tasks, for example.

The effects of continuous noise on performance has been well documented for loud noises (over 95dB(A)) but the effects appear to be similar for even quite moderate levels (75-85dB(A)). These effects are not only related to long-term exposure and the effects on hearing, but to performance efficiency on a variety of tasks (Smith, 1989). Although simple tasks (such as reacting to stimuli or clerical activities) seem to be unimpaired by noise, this is not true of more complex tasks such as multiple or continuous control tasks. The presence of noise in these latter tasks leads to increased errors, long reaction times and concentration on the main tasks at the expense of the others (Smith, 1989, cites Broadbent, 1979).

Smith (1989) notes that the effects of noise on performance are complicated and may be influenced by a variety of factors, some of which are not yet known. However he was able to draw some general conclusions, these are that noise (1) leads to the choice of certain strategies in preference to others; (2) often reinforces the use of the dominant strategy; and (3) impairs the control processes which track and change performance, which may make people rather inflexible and less adaptive to change. These three points have important implications for human tasks which typically demand that operators track events and adapt their strategies to suit the current status of the equipment with which they are working. It may be particularly important in operating equipment to be aware of developments; for example, an incident may start out in a familiar fashion, but may subsequently develop along unfamiliar lines. Thus the operator's initial strategy may be appropriate at the start of the task, but may become less appropriate as the incident develops. Noise from the warning system could hamper the ability of the operators to keep abreast of new developments and to modify their behaviour accordingly.

Conclusions and Organisation of the Book

Much can be done to improve auditory alarm displays, as current systems appear to be less than optimal. The human factors perspective has much to offer in the design, construction and implementing of such systems. The unique features of the auditory channel can be capitalised upon, rather than, as is typically the case, arbitrarily allocating warnings and signals to protect a particular device. Used appropriately, and perhaps combined with other media, auditory warnings and displays may be very effective. However care should be taken to consider the physical environment, the ambient noise levels and the task the operator is required to undertake. The auditory channel has an undoubted superiority in attention-getting ability, but care should be taken not to exploit this without good reason as this can lead to performance becoming unnecessarily impaired.

The chapters in this book represent many of the areas in which developments in auditory warning design, implementation and basic research have been taking place over the last few years, and as such present an up-to-date perspective on the whole issue of human factors in alarm design. The contributions range from those which focus entirely on auditory processes to those which take a broader perspective on human activity within a warnings-related environment. The nature of the focus of each chapter has determined the organisation of this book. Although a number of possible methods of organisation are possible, we have chosen to organise the book in the following way because it bears a relation to listening and responding to warnings in real time; thus it starts from auditory processes, moves on to psychological processes, and then on to broader, alarms-related issues. Many of the chapters cover both background theory and examples of application, but tend to take a different aspect of alarm design and implementation as their main thrust. In Part II, we have a sequence of three chapters concerned wholly with the first perceptual processes involved in alarm design, such as the issues of alarm localisation and detection. Important questions are 'Where are the sounds coming from?', 'Can they be heard?' and 'How can the design of alarm sounds be improved acoustically without necessarily altering their identity?' In this vein, Withington, in Chapter 2, deals with the crucial issue of alarm localisation with particular reference to emergency vehicles. She also touches on another issue which arises again later in the book, that of multimodal stimulation and the advantages that might be conferred by this approach. In Chapter 3, Robinson and Casali deal with the problems that arise when, as is often the case, hearing protectors are being worn when alarms are heard. This becomes an even greater problem when the hearer is suffering from some hearing loss, and this chapter looks specifically at this issue. In Chapter 4, Patterson and Datta stay with a purely acoustic and auditory perceptual approach, discussing how previously designed warning sets can be enhanced by applying specific acoustic techniques to already existing warnings.

In Part III, the authors start from the premise that acoustic issues are already dealt with and focus instead on the next stage of the process of listening to alarms; that is, the focus tends to be on questions such as 'What sorts of sounds should be designed?', 'Does urgency mapping have a role?' and 'How can the best designs for alarm sounds be elicited?' To this end, we have a sequence of four chapters exploring different, but related, aspects of auditory cognition. The first of these, Chapter 5, by Ballas, covers the theory as to how we listen to sounds in a broader context, and presents a general model of everyday sound perception. It also emphasises the importance of bearing in mind that when people are listening to alarm sounds, they are necessarily listening to other, usually more everyday, sounds as well. Chapter 6 (by Stanton and Edworthy) follows on from this more general thinking about cognitive processes and presents some empirical data obtained through applying a methodology proposed earlier (Stanton and Edworthy, 1994; Edworthy and Stanton, 1995), designed to generate user-centred alarm systems. Chapters 7 and 8 (by Haas and Edworthy, and Burt, Bartolome-Rull, Burdette and Comstock, respectively) focus more specifically on the issue of perceived urgency and urgency mapping in alarms. Chapter 7 contains an empirical study which looks at the relationship between psychoacoustic urgency and reaction time, and Chapter 8 looks more broadly at the relationship between psychoacoustic urgency and learned priority, as well as looking at some psychophysiological reactions to alarm sounds.

Part IV presents a series of chapters which again take a largely cognitive approach to looking at alarms but incorporate issues other than those pertaining directly to the design of the warning or alarm itself. Some of the chapters include other warning modalities, for example, which clearly have to be considered as soon as one looks beyond the auditory and cognitive issues to the more general, human factors, issues. In Chapter 9, Bliss considers the crucial issue of alarm mistrust and false alarms. It is important that any alarm, no matter how well it is designed, truly signals an alarm situation; otherwise, as Bliss shows, mistrust will begin to creep in and less than optimal performance will be achieved. Ongoing task performance is also considered in this chapter. Following on from this, Chapter 10, by Selcon, looks at some of the theoretical issues relating to the use of multimodal warnings, particularly perceptual integrality and the stroop effect. A military application is also described. The next chapter, by Rauterberg, continues the multi-modal theme, describing a pair of studies where the use of visual-only feedback was compared with the use of visual and auditory feedback. The approach to sound taken in this chapter is one which regards sounds as being signallers of events, and where the design of the sound is of secondary importance. In Chapter 12, Stanton and Baber explore the role of yet another type of alarm, the speech alarm. The chapter presents a study which compares text-based with speech-based alarms in a process-monitoring task. Mixed results are obtained and the study highlights the kinds of problems which can be encountered when trying to make objective and meaningful comparisons across different warning modalities.

Part V contains four chapters concerned either with charting problems which already exist in real working environments, or with proposing methods of improving some areas of practice. To some extent nearly all of the chapters of the book are driven by this same objective, but in these four chapters the applied focus is more central to the work reported. The first two chapters focus on aviation and the second two on the Intensive Care Unit (ICU). Chapter 13, by Noyes, Creswell Starr and Rankin, considers the complete warning system on flight decks of aircraft from a user's point of view, and presents data showing us how warnings and alarms are used and perceived in practice by a specific professional group, pilots. In Chapter 14, Hellier and Edworthy describe a pair of studies carried out with relation to a specific alarm design remit: that of constructing a new warning set based on an already established set so that the new set is both identifiable as being similar to the old set and yet different from it. Chapters 15 and 16 look at some of the issues currently causing concern in the ICU. Chapter 15, by Meredith, Edworthy and Rose, provides a 24-hour auditory 'picture' of alarm use in an ICU derived from video observation. In Chapter 16, by contrast, Welch provides a pair of empirical studies looking at medical staff's learning and retention of warnings, exploring the issue of urgency mapping in particular. In several instances these last chapters hark back to issues looked at from a slightly different perspective earlier in the book.

Acknowledgments

The authors would like to take this opportunity to thank Roy Patterson for allowing them to use Figure 1.1 from his work and Taylor and Francis (publishers) for permission to reproduce Figure 2.2.

References

Baber, C. (1991) *Speech Technology in Control Room Systems*, London: Ellis Horwood.

Blattner, M.M., Sumikawa, D.A. and Greenberg, R.M. (1989) 'Earcons and icons: their structure and common design principles', *Human–Computer Interaction*, **4**, 11–44.

Bliss, J.P., Gilson, R.D. and Deaton, J.E. (1995) 'Human probability matching behaviour in response to alarms of varying reliability', *Ergonomics*, **38** (11), 2300–2312.

Bregman, A.S. (1990) *Auditory scene analysis: the perceptual organisation of sound*, Cambridge, MA: MIT Press.

Edworthy, J. (1997) 'Noise and its effects on people: an overview', *International Journal of Environmental Studies*, **51**, 335–44.

Edworthy, J. and Adams, A.S. (1996) *Warning Design: A Research Prospective*, London: Taylor & Francis.

Edworthy, J. and Stanton, N.A. (1995) 'A user-centred approach to the design and evolution of auditory warning signals: 1. Methodology', *Ergonomics*, **38** (11), 2262–80.

Edworthy, J., Loxley, S. and Dennis, I. (1991) 'Improving auditory warning design: relationship between warning sound parameters and perceived urgency', *Human Factors*, **33**(2), 205–31.

Gaver, W.W. (1986) 'Auditory icons: using sound in the computer interface', *Human–Computer Interaction*, **2**, 167–77.

Hellier, E.J., Edworthy, J. and Dennis, I. (1993) 'Improving auditory warning design: quantifying and predicting the effects of different warning parameters on perceived urgency', *Human Factors*, **35**, 693–706.

Laroche, C., Tran Quoc, H., Hétu, R. and McDuff, S. (1991) '"Detectsound": a computerized model for predicting the detectability of warning signals in noisy environments', *Applied Acoustics*, **33**(3), 193–214.

Lazarus, H. and Höge, H. (1986) 'Industrial safety: Acoustic signals for danger situations in factories', *Applied Ergonomics*, **17**(1), 41–6.

Lees, F.P. (1974) 'Research on the process operator', in E. Edwards and F.P. Lees (eds), *The Human Operator in Process Control*, London: Taylor & Francis.

Marshall, E. and Baker, S. (1994) 'Alarms in nuclear power control rooms: current approaches and future design', in N.A. Stanton (ed.), *Human Factors in Alarm Design*, London: Taylor & Francis.

McClelland, I.L. (1980) 'Audible warning signals for police vehicles', *Applied Ergonomics*, **11**(3), 165–70.

Meredith, C. and Edworthy, J. (1994) 'Sources of confusion in intensive therapy unit alarms', in N.A. Stanton (ed.), *Human Factors in Alarm Design*, London: Taylor & Francis.

Patterson, R.D. (1974) 'Auditory filter shape', *Journal of the Acoustical Society of America*, **55**, 802–9.

Patterson, R.D. (1976) 'Auditory filter shapes derived with noise stimuli', *Journal of the Acoustical Society of America*, **59**, 640–54.

Patterson, R.D. (1982) 'Guidelines for Auditory Warning Systems on Civil Aircraft', CAA paper 82017, London: Civil Aviation Authority.

Patterson, R.D. (1985) 'Auditory warning systems for high-workload environments', in I.D. Brown, R. Goldsmith, K. Coombes, and M.A. Sinclair (eds), *Ergonomics International 85*, London: Taylor & Francis.

Patterson, R.D. (1989) 'Guidelines for the design of auditory warnings', *Proceedings of the Institute of Acoustics*, **11**(5), 17–24.

Patterson, R.D. (1990) 'Auditory warning sounds in the work environment', in D.E. Broadbent, J. Reason and A. Baddeley (eds), *Human Factors in Hazardous Situations*, Oxford: Clarendon Press.

Patterson, R.D. and Nimmo-Smith, I. (1980) 'Off-frequency listening and auditory filter asymmetry', *Journal of the Acoustical Society of America*, **67**, 229–45

Reinartz, S.J. and Reinartz, G. (1989) 'Analysis of team behaviour during simulated nuclear power plant incidents', in E.D. Megaw (ed.), *Contemporary Ergonomics*, London: Taylor & Francis.

Rouse, W.B. (1983) 'Models of human problem solving', *Automatica*, **19**, 613–25.

Sanders, M.S. and McCormick, E.J. (1987) *Human Factors in Engineering and Design*, New York: McGraw-Hill.

Smith, A. (1989) 'A review of the effects of noise on human performance', *Scandinavian Journal of Psychology*, **30**, 185–206.

Stanton, N.A. (1994) 'Alarm initiated activities', in N.A. Stanton (ed.), *Human Factors in Alarm Design*, London: Taylor & Francis.

Stanton, N.A. and Baber, C. (1995) 'Alarm-initiated activities: an analysis of alarm handling by operators using text-based alarm systems in supervisory control systems', *Ergonomics*, **38**(11), 2414–31.

Stanton, N.A. and Edworthy, J. (1994) 'Towards a methodology for designing representational auditory alarm displays', in S. Robertson (ed.), *Contemporary Ergonomics 1994*, London: Taylor & Francis.

Stokes, A., Wickens, C.D. and Kite, K. (1990) *Display Technology: human factors concepts*, Warrendale, PA: Society of Automotive Engineers, Inc.

Wickens, C.D. (1984) *Engineering Psychology and Human Performance*, Columbus, OH: Merrill.

Wickens, C.D. (1990) *Engineering Psychology & Human Performance*, New York: Harper Collins.

Zwaga, H.J.Z. and Hoonhout, H.C.M. (1994) 'Supervisory control behaviour and the implementation of alarms in process control', in N.A. Stanton (ed.), *Human Factors in Alarm Design*, London: Taylor & Francis.

Zwicker, E. and Scharf, B. (1965) 'A model of loudness summation', *Psychological Review*, **72**, 3–26.

PART II
ACOUSTICS AND
AUDITORY PROCESSES

2 Localisable Alarms

DEBORAH J. WITHINGTON, *University of Leeds*

Introduction

The ability to localise a sound source is an evolutionary prerequisite for animals' (including humans) survival. For example, when hearing the crack of a twig, as a predator approaches, there is simply not time to wait and look around to check where the sound is coming from. To survive an animal must react instantly, as soon as the audible signal is received. Similarly, for predators, a rustle of leaves may indicate where their potential prey is hiding, and locating that position will determine whether or not they eat on that occasion. It is, therefore, safe to conclude that pinpointing sound is something we do well. In reality we can localise a sound to an accuracy of about five degrees, given the right type of sound (Makous and Middlebrooks, 1990). This level of accuracy is less than that for visual spatial acuity, but more than adequate for survival purposes.

There is one particular part of our central nervous system that plays a vital role in the detecting of and, equally importantly, the response to a sound source. This area is part of the mid-brain and is called the superior colliculus (SC) (Stein and Meredith, 1993). The SC also responds to light and touch (for example, King, 1990). The implications of this multimodality will be discussed later in the chapter with reference to combining light and sound in emergency egress systems. Events that happen in the SC all happen subconsciously and, depending upon the type of stimulus, will elicit an orienting reaction either to or away from the stimulus source (Sahibzada *et al.*, 1986).

Neurophysiologists studying the properties of neurones in the SC together with psychoacousticians studying human responses to sound have enabled us to understand how the brain processes information relating to a sound source and, importantly, what type of sound is needed for a degree of accuracy to be achieved. It has long been recognised that localising a sound source requires a vast amount of neural processing (Knudsen *et al.*, 1987). Only certain types of sounds are inherently localisable and what is crucial is that they contain a large spectrum of frequencies, that is broadband noise. Pure tones, simple tone combinations or narrowband noise cannot be localised. To understand why this is the case, the cues given by sound, that the brain can recognise, must be considered.

We can hear a vast range of frequencies, from approximately 20Hz to 20kHz, although this range diminishes as we age. There are three main types of information that allow the brain to localise sound. The first two are known as binaural cues because they make use of the fact that we have two ears, separated by the width of our head. A sound which emanates from either side of the mid-line will arrive first at the ear closest to it and will also be loudest at the ear closest to it. At low frequencies the brain recognises differences in the time of arrival of the sound between the ears, and at higher frequencies the salient cue is the loudness/intensity difference between the sound at each ear. The use of these two types of cue is known as the 'duplex' theory and was proposed by Lord Raleigh as long ago as 1877.

For single frequencies these cues are, however, spatially ambiguous. The inherent ambiguity has been described as the 'cone of confusion' and this arises from the fact that for any given frequency there are numerous spatial positions that generate identical timing/intensity differences and these can be graphically represented in the form of a cone, the apex of which is at the level of the external ear. The cone of confusion is the main reason for our not being able to localise pure tones (Blauert, 1997; Wightman and Kistler, 1993).

The final main piece of information processed by the brain regarding sound localisation is called the head-related transfer function (HRTF) (Carlile and King, 1993). The HRTF refers to the effect the external ear has on sound. As a result of passing over the bumps or convolutions of the pinna, the sound is modified so that some frequencies are attenuated and others are amplified. Although there are certain generalities in the way the sound is modified by the pinnae, the HRTF of any one person is unique to that individual. The role of the HRTF is particularly important when we are trying to determine whether a sound is immediately in front of, or directly behind, us. In this instance the timing and intensity differences are negligible and there is consequently very little information available to the central nervous system on which to base a decision of 'in front' or 'behind'. So, to locate the direction of a sound source, the larger the frequency content, to overcome the ambiguities inherent to single tone sounds, the better the accuracy.

Knowing that we need a multi-frequency sound for localisation, how can this be combined with our need to use sound as an alarm? There are a myriad of different uses for sound in alarms and in some cases the addition of a localisability component would be superfluous. However, there are other alarms in which the lack of localisability/directionality is potentially highly dangerous. An excellent example of the latter is the siren used by emergency vehicles (Withington and Chapman, 1996). An everyday occurrence for the majority of drivers is the sound of an emergency vehicle siren, whether from an ambulance, police car or fire appliance. When the emergency siren is heard, drivers look all around, trying to determine from which direction the sounds are coming. The visual cue is required because the sound alone gives no clue as to which direction the vehicle is coming from. The driver is not able to take appropriate avoiding action until the emergency vehicle is seen, often too late to allow a clear path to be created for the emergency vehicle. The uncertainty regarding the direction of approach costs lives. Any improvement in the sound quality of the siren, which enables road users to take earlier evading action, would both reduce the journey time and enhance the safety for emergency vehicles attending emergencies and thus strengthen the service provided. It would also be safer for road users, pedestrians and drivers alike.

Difficulties in determining the direction from which emergency vehicle sirens are approaching are widely acknowledged. In fact the emergency vehicle siren has been described as 'an extremely limited audible warning device' (De Lorenzo and Eilers, 1991). A recent study in the *Annals of Emergency Medicine* has shown that an ambulance is most susceptible to collisions with other vehicles when crossing road junctions. This happens, primarily, because the drivers of the cars or trucks are unable to determine accurately the direction of the approaching ambulance. In one year, in the USA alone, 537 injuries and 62 deaths arose from accidents involving ambulances (Hunt *et al.*, 1995).

So why do they not work? Simply, because the frequency content of the siren sounds is so poor. Typically emergency vehicle siren sounds are emitted over the frequency range 500Hz–1.8kHz, far too narrow a frequency range for localisation. Although the anecdotal evidence is plentiful, there has not been, until recently, a thorough study into the localisability of emergency vehicle sirens. The lack of research has been due, primarily, to the difficult nature of tackling such a concept in a meaningful way. It would be of little value, for example, to test siren sounds by asking people seated in a room, or in the open, where they thought the sound originated. The 'realism element' was overcome in a recent study at Leeds University in which participants 'drove' in a simulator (Withington, 1996).

Experimental Study

Approximately 200 participants (age range 19 to 57 years) were tested. They all had previous driving experience. A hearing test, in the form of an audiogram, which tested the ability to hear a range of frequencies, was performed on most participants. Eight loudspeakers were positioned at 45 degree intervals around the car's azimuthal plane. The speakers were not visible to the participants, and the sound from each speaker was intensity and spectrality matched at a position which marked the driver's head within the car. Matching each speaker in the way described resulted in the only variable each time the sound was played being that of direction. A response panel was mounted by the steering wheel in the car. There were 16 positions marked on the panel which represented the horizontal plane around the car. This allowed greater accuracy in determining the errors made by the participants than just having eight response buttons that corresponded directly to the number of speakers. On hearing a siren sound noise, participant drivers pressed the response button that they judged as equating to the location from which the sound originated. The sound source location was varied randomly, with the constraint that in each session all eight speaker positions were activated twice. The participants began trials after a period of familiarisation with the car (a 10-minute 'drive' along a preset rural route). During the trials the participants were asked to maintain a speed of 40 mph which the experimenter monitored continuously.

Four existing sirens were tested. These were the 'hilo' siren characterised by a two-tone sound (670–1100Hz, 55 cycles/min); the 'Pulsar', a pulsating sound (500–1800Hz, 70 cycles/min.); the 'wail', a continuous sound rising and falling (500–1800Hz, 11 cycles/min) and the 'yelp', a continuous and fast warbling sound (500–1800Hz, 55 cycles/min). In each trial the same siren sound was used. The siren used was delivered at or around road junctions on the test track.

Data were gathered showing conclusively the poor localisation characteristics of existing sirens, though some were better than others; for example, the traditional 'hilo' siren was significantly worse for localisation than the 'yelp'. Nevertheless even the best of the current sirens was associated with exceedingly poor front/back accuracy: participants had difficulty identifying whether the sound emanated from the front or rear of the car, to the extent that they were wrong more often than not. It is worth noting that, if participants just guessed front or back, in theory they would have been correct 50 per cent of the time.

In the aforementioned study, in addition to testing the existing siren sounds, a range of new sounds optimised for both alerting (Patterson, 1982, Haas and Edworthy, 1996) and localisation were also tested. The data obtained from the simulator studies using the new siren sounds, characterised by pulses of rapidly rising frequency sweeps followed by a burst of broadband noise, indicated a much better localisation than with the old sirens. With the front/back decision, participants got it right 82 per cent of the time (whereas, with the old siren, front/back was incorrectly judged 56 per cent of the time), and left/right accuracy was increased to 97 per cent (from 79 per cent).

Following the simulator trials it was necessary to test the new siren sounds in real emergency situations. The new sounds were used in trials with ambulances in West Yorkshire and London, fire appliances in Leicestershire and a police pursuit car in West Yorkshire.

Road Trials

The road trials produced data which complemented the simulator trials (Withington and Paterson, 1998). Using an on-board video camera, which recorded the view of the road ahead of the emergency vehicle, road user responses to both the old siren and the new sirens were compared. As a result of discussions with the drivers of the emergency vehicles, various road user behaviours were searched for, such as other road users failing to indicate their proposed direction of manoeuvre. The data showed that with the new siren ambulance lane changes were reduced threefold, so making journeys smoother and faster. Instant recognition of the direction of the ambulance's approach increased by nearly 25 per cent, which meant that other drivers cooperated more quickly and more effectively with the ambulance's passage. Clearer, well-signalled manoeuvres by road users were more achievable and increased by 17 per cent. Finally journey times were reduced by up to 10 per cent. This reduction in journey time was not due to the fact that the emergency vehicle was travelling faster and, often, owing to the natural caution of driving with something new, their speed was actually less. It was due, rather, to the fact that the journey was made easier by road users reacting more quickly, and more appropriately, to the new siren, allowing a clearer path through the traffic for the emergency vehicle.

It seems, therefore, that the two crucial requirements of a siren sound, to alert road users to the presence of an emergency vehicle and to inform the listeners of the direction of approach, can be met by utilising our scientific knowledge of the best types of sounds to fulfil these criteria, thus resulting in a more effective sound signal.

Other Applications

There are many other cases in which changes in the nature of the sound would enhance the efficacy of the application. For example, distress signal units (DSUs), personal alarms worn by firefighters, are activated under conditions of extreme urgency. The sound emitted should not only alert colleagues of the stricken firefighter, but should also provide useful information to aid their search for him or her. Existing DSUs use single-tone, relatively high-frequency signals of approximately 3kHz. These sounds are not ideal in terms of perceived urgency and, more importantly, they are impossible to localise. Localisation errors are, in fact, greatest for sounds around 3kHz (Stevens & Newman, 1936). Research is currently under way at the University of Leeds to improve the sounds emitted by DSUs.

It is also worthwhile considering applications in which the addition of a sound would enhance the function. An example of this is the addition of sound to emergency egress lighting systems. Effective emergency egress is of paramount importance for individuals whether it be from aeroplanes, hotels, industrial complexes or ferries and even in the home environment. It is surprising, therefore, that the locations of emergency exits and/or escape routes are indicated solely by visual means. Any auditory information is provided in the guise of an 'alarm' which merely alerts people to the imminent danger.

It could be argued that we are, primarily, visual creatures, and there is no doubt that we have a phenomenal accuracy when pinpointing a spatial visual signal (to a fraction of a degree); hence visual instruction for emergency egress should suffice. There are two arguments that can be used against the contention that visual signs alone are optimal for egress. Firstly, research has shown that the area of the brain that responds to spatial sensory information (the superior colliculus), which also initiates the response to the sensory stimulus, contains cells that respond to more than one sensory modality. These neurones respond to light by itself and also to sound alone. However, when light and sound are presented together, the response of these cells is far greater than the summation of the response to either modality alone (Stein and Meredith, 1993). In other words, to activate an individual to 'flight', a stimulus containing both light and sound will be far more effective than light or sound alone. Secondly, and most importantly, there are many situations in which our visual ability can be hampered, for example in the case of smoke or chemical fumes, or as a result of an inherent visual disability (Rutherford, 1997). In these instances it is imperative that an alternative modality be activated, and the use of sound is the obvious solution.

It is important to note that the combination of some types of sound with emergency lighting would do more harm than good. Sounds with poor frequency content, and even instructions conveyed by speech (which is in effect narrowband noise), are inappropriate for emergency egress as they do not convey 'directional' information.

Tests at Leeds University in a smoke-filled environment, filmed by a thermal imaging camera, showed that it took a subject 3 minutes 50 seconds to find a conventional exit sign, relying predominantly on their memory of the immediate environment. In contrast, when an appropriate sound was added to the standard sign the same individual took 15 seconds to find their way out of the hazardous environment.

Conclusions

Sound is incredibly versatile as a means of locating emergency egress routes. Varying the pattern of noise can draw people along complex routes. Furthermore addition of melodic complexes can provide 'up' or 'down' information which could be placed at strategic stairwells on an exit route.

Finally it is worthwhile noting that in control rooms which may house vast consoles, including many visual and auditory signals to alert or warn the operators of a myriad of problems, the use of pure tone alarms may, by their very nature of inherent spatial ambiguity, cause unnecessary confusion for the operator. Furthermore, in the quest to reduce the number of signals in such busy environments, it may be possible to include directional characteristics in such warning alarms so that optimal visual and auditory alarms are created.

References

Blauert, J. (1997) *Spatial Hearing*, Cambridge, MA: MIT Press.

Carlile, S. and King, A.J. (1993) 'Auditory Neuroscience: from outer ear to virtual space', *Current Biology*, **3**, 446–8.

De Lorenzo, R.A. and Eilers, M.A. (1991), 'Lights and Siren: A review of emergency vehicle warning systems', *Annals of Emergency Medicine*, **20**, 1331–5.

Haas, E.C. and Edworthy, J. (1996) 'Designing urgency into auditory warnings using pitch, speed and loudness', *Computing and Control Engineering Journal*, August, 193–8.

Hunt, C.R., Brown, L.H., Cabinum, E.S., Whitely, T.W., Prasad, N.H., Owens, C.F. and Mayo, C.E. (1995) 'Is ambulance transport time with lights and siren faster than without?', *Annals of Emergency Medicine*, **20**, 507–11.

King, A.J. (1990) 'The integration of visual and auditory spatial information in the brain', in D.M. Gutherie (ed.), *Higher Order Sensory Processing*, Manchester: Manchester University Press.

Knudsen, E.I., du Lac, S. and Esterley, S.D. (1987) 'Computational maps in the brain', *Annual Review of Neuroscience*, **10**, 41–65.

Makous, J.C. and Middlebrooks, J.C. (1990) 'Two-dimensional sound localization by human listeners', *Journal of the Acoustical Society of America*, **87**, 2188–200.

Patterson, R.D. (1982) 'Guidelines for Auditory Warning Systems on Civil Aircraft', CAA paper 82017, London: Civil Aviation Authority.

Rutherford, P. (1997) 'Auditory navigation and the escape from smoke filled buildings' in R. Junge (ed.), *CAAD Futures*, Proceedings of the 7th International Conference on Computer Aided Architectural Design Features, Dordrecht: Kluwer Academic Publishers.

Sahibzada, N., Dean, P. and Redgrave, P. (1986) 'Movements resembling orientation or avoidance elicited by electrical stimulation of the superior colliculus in rats, *Journal of Neuroscience*, **6**, 723–33.

Stein, B.E. and Meredith, M.A. (1993) *The Merging of the Senses*, Cambridge, MA: MIT Press.

Stevens, S.S. and Newman, E.B. (1936) 'The localization of actual sources of sound', *American Journal of Psychology*, **48**, 297–306.

Wightman, F.L. and Kistler, D.J. (1993) 'Sound Localization', in W.A. Yost, A.N. Popper and R.R. Fay (eds), *Human Psychophysics*, New York: Springer-Verlag.

Withington, D.J. (1996) 'The quest for better ambulance sirens', *Ambulance UK*, **11**, 20–21.

Withington, D.J. and Chapman, A.C. (1996) 'Where's that siren?', *Science and Public Affairs*, Summer, 59–61.

Withington, D.J. and Paterson, S.E. (1998) 'Safer Sirens', *Fire Engineers Journal*, **48**, 6–10.

3 Audibility of Reverse Alarms under Hearing Protectors and its Prediction for Normal and Hearing-impaired Listeners

GARY S. ROBINSON and JOHN G. CASALI,
Virginia Tech

Introduction

It has been estimated that in the USA as many as nine million workers are exposed to occupational noise levels exceeding a time-weighted average (TWA) of 85dB(A) for an eight hour day (EPA, 1981). Exposures of this magnitude are sufficient to cause permanent hearing loss. Because of their convenience and relatively low cost, hearing protection devices (HPDs) are the most widely implemented solution to the problem of employee noise exposure in industry today. However, individuals who work in high-noise environments often complain that wearing HPDs interferes with their ability to hear auditory warning and indicator sounds (Suter, 1989).

Few studies have addressed the detectability of auditory warning signals presented in noise while wearing hearing protection. Fewer still have included hearing-impaired listeners. Usually only the occluded and unoccluded masked thresholds are obtained and compared. However a few researchers (Coleman *et al.*, 1984; Laroche *et al.*, 1991; Lazarus, 1980) have suggested methods for predicting the occluded masked threshold of an auditory warning or alarm.

Detecting Signals in Noise

Perhaps the most methodical examination of the problems associated with the audibility of alarms or warnings in noise while wearing HPDs was conducted by Wilkins and Martin (1977, 1981, 1982, 1984, 1985). The first two experiments (1977, 1982) determined masked thresholds of normal-hearing subjects with and without use of HPDs for six common warning sounds presented against various background noises. Results showed significantly lower masked thresholds, indicating an improved ability to detect the signals in the occluded condition (wearing hearing protection), for several of the sounds in noise levels of 90 and 95dB(C).

Wilkins and Martin (1981, 1984, 1985) also performed a series of experiments aimed at determining how inattention affects the detection of warning signals, if various noise and signal parameters interact during periods of inattention, and if these interactions affect masked occluded thresholds. No difference in detection performance between vigil (subjects intentionally listened for the signals) and loaded conditions (subjects performed a secondary task) was found. This finding contradicts results of an experiment reported by Fidell (1978) in which detection thresholds were found to be higher than predicted when a loading task was performed. Wilkins and Martin (1981, 1982) suggest that the different results may have been due to procedural artefacts. If so, the increased detection thresholds measured in the Fidell (1978) study may have been due in part to shifts in the subjects' criteria rather than a change in their sensitivities. Finally Wilkins and Martin (1984, 1985) found that a warning signal's contrasts with both noise and any irrelevant sounds are important parameters in a detection task, with the signal's contrast with irrelevant stimuli being slightly more important than its contrast with the background noise.

Wilkins (1984) also conducted a field study in a manufacturing facility intended to investigate the interaction of hearing loss and HPDs with warning signal detectability. Neither normal-hearing nor hearing-impaired listeners showed any significant differences in detection of the alarm between the occluded and unoccluded conditions. However the author cautioned readers about taking the results too literally and cited several possible confounding variables in his study, including the inability to control the background noise levels and thus the signal-to-noise ratios.

When investigating the detectability of pure tones against an 88dB(A) background noise, Forshaw (1977) found small but statistically non-significant differences between the occluded and unoccluded thresholds for normal-hearing subjects, with the masked occluded thresholds being slightly lower. A single hearing-impaired listener was said to have shown no adverse effect when wearing the HPD, but was also said to have been completely unable to detect a 3000Hz tone while wearing the HPD.

Abel *et al.* (1985) conducted an experiment which included both normal and hearing-impaired listeners. Results indicated that, for normal-hearing subjects, use of a foam earplug reduced the masked threshold of a third-octave band of noise centred at 3000Hz by 3–6dB in 84dB(A) noise, but did not significantly affect the detectability of a third-octave band of noise centred at 1000Hz. The hearing-impaired listeners showed significant increases in the masked threshold, indicating a reduced ability to detect the signals for the signal centred at 3000Hz when the earplug was worn, but no significant changes in masked thresholds were found for the signal centred at 1000Hz.

Predicting the Audibility of a Signal in Noise

Several researchers have attempted to develop models that can be used to predict masked occluded thresholds for normal and hearing-impaired listeners. Coleman *et al.* (1984) developed a model based on the auditory filter work of Patterson (1974, 1976) and Patterson *et al.* (1982), then conducted a series of experiments to validate the model. Predicted occluded masked thresholds consistently overestimated the mean measured thresholds by about 5dB (erring on the conservative side). The rather complex model requires knowledge of the spectral make-up of the masking noise, the spectral attenuation characteristics of the HPD, the spectral characteristics of the signal, the pure-tone threshold of the individual (or group) in question and an estimate of the auditory filter width of the individual (or a suitable population estimate).

Lazarus (1980) also proposed a method for determining whether or not a signal would be audible to an individual wearing an HPD in noise. The method requires knowledge of the third-octave spectral characteristics of the noise, the spectral attenuation characteristics of the HPD and calculation of a masked threshold for the signal under both the occluded and unoccluded conditions. However, a recommendation as to how these masked thresholds are to be calculated was not suggested. No empirical evidence of the accuracy of this method was reported.

Laroche *et al.* (1991) have developed a computer program capable of predicting masked thresholds in noise (both with and without hearing protectors) which takes into account hearing loss due to age. The model does not, however, consider the effects of noise-induced hearing loss or hearing loss due to injury, disease or other etiology. The authors do acknowledge the model's shortcomings and discuss plans for expanding the model, not only to include hearing loss due to other etiologies, but also to include the effects of sound propagation in sound fields with vastly different reverberation characteristics, and variation of the spectral and temporal characteristics of both the background noise and the warning signal (the current model assumes steady-state conditions).

Finally, ISO (1986) presents a simplified method for calculating a masked threshold for auditory alarms and warnings in noise. Two methods are presented, one requiring the one-third octave-band spectrum of the noise, the other requiring the octave-band noise spectrum. Although HPD use can be accounted for if a reliable estimate of the spectral attenuation characteristics of the HPD are available, the method presented does not take the hearing threshold of the listener into account.

Summary

On the basis of the preceding discussion, it appears that, for normal-hearing individuals, the use of HPDs will not adversely affect their ability to detect warning or indicator sounds in high-noise environments above about 85dB(A). In fact their use may actually improve signal audibility in some circumstances. The explanation for such an effect is that, although the HPD attenuates the signal and the noise equally, it reduces their levels to the point where the cochlear distortion present at high noise levels is reduced or eliminated, thus enabling the ear to better distinguish the signal from the noise (Suter, 1989). This effect, however, is likely to be HPD-specific in that the attenuation of the device must be sufficient to reduce the distortion present in the cochlea.

For hearing-impaired listeners, particularly those with high-frequency neural loss, it appears that use of HPDs in noise may impair the ability to detect warning signals. The effects, however, may be limited to frequencies above 1000Hz (assuming noise-induced hearing loss). Similar conclusions were reached by Wilkins and Martin (1987). The explanation for this effect is that the HPD attenuates the signal (and also the noise) to such an extent that the sound reaching the ear is below the already compromised auditory threshold of the user (Abel *et al.*, 1985; Lazarus, 1980). However the point at which performance begins to degrade is not known. This question provided impetus for the study described herein. As many as 28 million Americans exhibit significant hearing loss due to a variety of etiologies, such as pathology of the ear, ototoxic drugs and hereditary tendencies. Of these, over 10 million have losses which are directly attributable to noise exposure (NIH, 1990). Clearly, the ability of hearing-impaired individuals to hear alarms or warnings while wearing HPDs is an important safety issue that affects a large segment of the industrial population.

Although there is some evidence to the contrary, it would appear that, for well-designed warning signals, inattention may not significantly affect the detection of the signals in noise when wearing HPDs. However, for this to be true, the signal must be distinct from both the background noise and any incidental sounds which may occur. Furthermore, although methods have been developed to predict the occluded masked threshold of auditory alarms and warnings, they are complex and do not readily lend themselves to general use in industry, since they require information (individual auditory filter widths) that may not be generally available or easily obtained by persons responsible for administering hearing conservation programmes. Therefore it is doubtful whether these prediction methods will find general acceptance in the near future. Of the methods mentioned in the preceding discussion, the ISO standard procedures are the easiest to implement.

Objective

The objective of the research effort described herein was to determine how the ability of normal-hearing and hearing-impaired listeners to detect an audible alarm in noise while wearing an earmuff changed as the noise level and the signal-to-noise ratio was systematically varied. The experimental paradigm was based on the theory of signal detection (TSD; see Green and Swets, 1988).

The experiment investigated the detectability of signals in noise only for the occluded (wearing an HPD) condition. The experiment was not intended to determine the difference in signal detectability between the occluded and unoccluded conditions, but rather to determine how the ability to hear an alarm varies with hearing level, noise level and signal-to-noise ratio only in those environments in which the use of HPDs is required.

Method

HPD Selection

The decision to use an earmuff was made since, on the evidence of literature review, it is believed that an earmuff represents a 'worst-case' scenario, for two reasons. First, earmuffs generally show less attenuation than earplugs at low frequencies, so there is greater opportunity for upward spread of masking to occur beneath the earcups, thereby reducing the audibility of a signal. Second, since earmuffs typically exhibit slightly better attenuation at frequencies from 1000 to 4000Hz (the frequency range of maximum human sensitivity) than do many premoulded earplugs, an earmuff would then likely directly attenuate warning signals slightly more than an earplug, possibly to a point below the threshold of hearing for some hearing-impaired individuals. Therefore the experimental results should be slightly conservative since, if a sound is audible while wearing an earmuff, it should also be audible when using an earplug, but the converse would probably not be true. Similar sentiments were expressed by Lazarus (1980).

A Bilsom Viking earmuff, manufactured in Sweden by Bilsom International, was the device chosen for use in the experiment. This large-volume, high-attenuation earmuff (having a noise reduction rating (NRR) of 29 at the time of the study) was identified by the research sponsor as an appropriate earmuff for use in the noise levels being investigated (85 to 95dB(A)). It is a heavy-duty product that is widely used in US (as Model 2318) and European (as Model 2421) industry (D. Weeks, personal communication, 3 October 1994). In addition, since the experimenter had considerable prior experience with the device, having found that consistent fits across sessions with the same subject as well as across subjects were easily obtained, it was believed that differences in signal detection due to HPD fitting problems would be minimised. The manufacturer-supplied attenuation data as well as the one-third octave band spectral attenuation characteristics of the Bilsom Viking earmuff are shown in Table 3.1.

Background Noise Spectra

Pink noise (flat by octaves) was used as the background noise in the experiment. Use of pink noise provided the greatest opportunity for upward spread of masking to decrease the audibility of the signal, but less opportunity for direct masking than would a mid-range-biased noise. Also interaction of the pink noise with the relatively poor low-frequency attenuation characteristics inherent with earmuffs was expected to further reduce the audibility of the signal at all noise levels. Pink noise is also a popular 'generic' noise used in psychoacoustic studies which have industrial workplace implications.

Warning Signal

The warning signal used in the study was a standard reverse alarm (manufactured by Caterpillar, Inc., PN 3T-1815) commonly found on heavy equipment and meeting the requirements for a Type A device as described in SAE J994b-1978 (SAE, 1978). This type of warning signal was identified as one of the more common alarm/warning signals across all of the research sponsor's manufacturing facilities, considered to be representative of industrial plants which rely heavily on diesel-powered vehicles. The spectrum of the particular reverse alarm (as measured in the Auditory Systems Laboratory's anechoic room) is illustrated in Figure 3.1.

Table 3.1 Spectral attenuation characteristics of the Bilsom Viking earmuff

One-third octave band centre freq. (Hz)	Measured attenuation (dB)	Manufacturer's data (circa 10/93)	
		Attenuation (dB)	Standard deviation (dB)
100	9.2	23.0	2.2
125	10.0		
160	12.0		
200	16.4		
250	20.8	25.0	2.0
315	23.0		
400	27.4		
500	32.0	31.0	1.9
630	36.4		
800	41.3		
1 000	44.1	36.0	2.2
1 250	42.8		
1 600	42.3		
2 000	41.0	40.0	2.0
2 500	38.8		
3 150	43.2	42.0	2.3
4 000	48.5	42.0	2.3
5 000	46.5		
6 300	43.2	39.0	3.2
8 000	43.4	37.0	3.0

Source: Casali *et al.* (1995) and product packaging (circa October 1993).

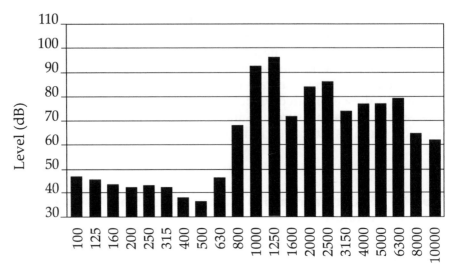

One-third octave band centre frequency (Hz)

Figure 3.1   Reverse alarm spectrum (Caterpillar, Inc., PN 3T-1815)

As depicted in Figure 3.1, the alarm has most of its energy in the bands centred at 1000 and 1250Hz, with fairly strong harmonics present in the bands centred at 2000 and 2500Hz. As shown in the figure, the levels in these four one-third octave bands are 93.5, 96.5, 84.5 and 86.5dB, respectively. For all practical purposes, the components in all other bands are unimportant since their levels are sufficiently below the levels in these four bands as to be negligible. The alarm operates with a 1s period and a 50 per cent duty cycle. These characteristics are in keeping with current warning signal design standards, including ISO 7731-1986 (ISO, 1986).

Subjects

A total of 12 subjects, ranging in age from 18 to 73 years, participated in the experiment as paid volunteers. Screening criteria were based on the subject's pure-tone hearing threshold. Since hearing level was a manipulated factor in the experiment, a detailed discussion of the screening criteria appears in the experimental design section later in the chapter.

Facilities and Instrumentation

All experimentation was conducted in the Auditory Systems

Laboratory on the Virginia Tech campus. This laboratory contains two audiometric test chambers, a reverberant room and an anechoic room, as well as a variety of support equipment and instrumentation. The laboratory itself is located in an acoustically isolated area so as to maintain a quiet environment for testing purposes.

The anechoic facility, a modified Eckel Corporation anechoic chamber, was used for all audiometric tests, spectral analysis and digital sampling of the reverse alarm, and matching the spectral output of the digitised signal with that of the original alarm. The reverberant room, an extensively modified Industrial Acoustics Corporation (IAC) audiometric test booth, was used for all experimental sessions. This was done to approximate the near-diffuse sound field conditions often encountered inside large industrial plants with reflective wall, floor and workstation surfaces.

A schematic diagram of the experimental apparatus is shown in Figure 3.2. Presentation of all test stimuli (signals and noise) and recording of all subject response data were performed using a Macintosh IIci microcomputer. A Larson•Davis (L•D) 3100D real-time analyser served as the pink noise generator and was controlled via computer. The pink noise output of the L•D 3100D RTA was directed to a Scott Model 458A (65w/ch) integrated audio amplifier and a Realistic Model 31-2000A octave band equaliser, used to shape the noise. The noise output of the Scott amplifier was directed to a pair of Infinity RS6b three-way loudspeakers situated inside the reverberant room, as shown in Figure 3.2 (speakers 1 and 2). Tests of this arrangement using pink noise showed the sound field about the subject's head centre position to be uniform and non-directional.

The digitised warning signal was presented via the computer's digital audio output. The signal was shaped via an AudioControl octave band equaliser and a Ross R31M third-octave band equaliser and amplified using an Adcom GFP-545II (100 w/ch) and GFP-555II amplifier/pre-amplifier combination. Output from the Adcom amplifier was directed to a single Klipsch K57K mid-range horn driver located behind the subject, just to the right of the door (speaker 3). The loudspeaker was positioned in this manner in an attempt to represent what is believed to be the 'worst-case' scenario for such an alarm, the situation in which a vehicle is approaching an unsuspecting subject from behind. This set-up provided a faithful reproduction of the acoustic characteristics of the original reverse alarm and allowed the presentation level of the alarm to be adjusted in order to achieve the various signal-to-noise ratios used in the experiment, since the output of the actual reverse alarm could not be varied.

A 20-inch (diagonal) computer monitor was used to present visual information, instructions and feedback to the subject during the

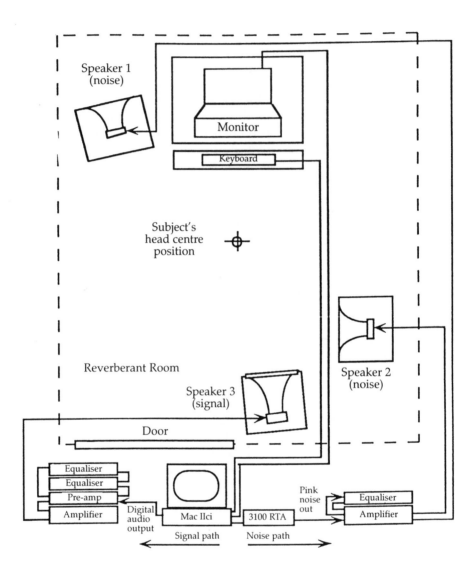

Figure 3.2   Schematic diagram of the experimental apparatus

course of the experiment. Subject responses were made using a modified computer keyboard and were monitored on a 12-inch monochrome display located at the experimenter's station.

The ambient noise levels in the reverberant room are 20dB or less in the octave bands centred at 125 and 250Hz, and less than 9dB in the octave bands centred from 500 to 8000Hz (Casali and Robinson, 1990). Reverberation times (RT_{60}) in the test space at the third-octave bands from 125 to 8000Hz ranged from 0.51 to 1.2s. The octave band spectrum of the pink noise was flat (to within 2.1dB) from 63 to 8000Hz.

Experimental Design

The experimental design used in the research effort described herein was a mixed three-factor design. The data were analysed using repeated-measures analysis of variance (ANOVA). The three independent variables represented in the design were hearing level (HL), noise level (NL) and signal-to-noise ratio (S/N). Hearing level was the single between-subjects variable and had three levels (normal hearing, slight loss and mild-to-marked loss). The descriptive terms used are in general agreement with those used by Miller and Wilber (1991) when related to the hearing levels represented by each category described below.

Subjects were screened on the basis of their pure-tone average (PTA) hearing levels over the frequency range from 500 to 2000Hz. This was done since the reverse alarm used in the experiment had most of its energy contained in the 1000 to 2000Hz range and it was believed that the frequencies above 2000Hz would have little impact on the detectability of the signal. Normal-hearing subjects were defined as individuals whose PTA hearing levels in both ears in the frequency range of interest was between 0 and 20dBHL. Subjects falling into the second group (slight loss) were required to have PTA hearing levels in both ears between 20 and 40dBHL. Subjects whose PTA hearing levels in both ears were above 40dBHL but below approximately 60dBHL qualified for the third group. An additional requirement placed on all subjects was that the range of hearing in either ear across the frequencies of interest (500 to 2000Hz) could not vary by more than 30dB. Five subjects each were recruited for groups one and three. However, only two subjects meeting the criteria for the middle group could be recruited.

Noise level was a within-subject variable with three levels: 85, 90 and 95dB(A). This range of noise levels encompasses over 90 per cent of the levels commonly encountered in industry (EPA, 1981). Signal-to-noise ratio was a within-subject variable with four levels (0, –8, –16 and –24dB). This variable represents the broadband A-weighted sound pressure level of the signal relative to that of the noise. The levels chosen for the signal were based on pilot tests which indicated that the masked threshold for the signal in pink noise for normal-hearing listeners was in the vicinity of –20 to –25dB.

Each subject participated in 12 experimental conditions (four signal-to-noise ratios at each of three noise levels). Presentation order of conditions for each subject was accomplished by way of random assignment. In addition, four earmuffs were randomly assigned to each condition so that each subject would use each earmuff three times during the course of the experiment.

The dependent measure used in the experiment was the proportion of the area under the receiver operating characteristic (ROC) curve, P(A), as recommended by Swets (1986, 1988). One advantage to using P(A) is that it is independent of the underlying distributions governing the subjects' responses (Robinson & Watson, 1972). In addition, since values of P(A) range from 0.50 for chance performance to 1.0 for perfect detection (Swets, 1986), the measure may be thought of as the *accuracy* with which a subject can discriminate a signal from noise (Swets, 1988, Swets *et al.*, 1979). Alternatively, a value of P(A) of 0.75 may be thought of as an indication of 'threshold' since it represents a level of performance halfway between chance and perfect performance.

Experimental Procedure

The experiment was conducted using the rating procedure methodology of signal detection theory. Signal detection theory was chosen since it was felt that, if one of the classical psychophysical methodologies had been used, the data might have been biased by one or more non-experimental factors (inattention, motivation, anxiety, experience, and so on). The rating procedure was selected over the other methodologies for its economy, since only one experimental session is required for each experimental condition.

As stated earlier, each subject was required to attend 12 experimental sessions in addition to the screening session and one practice session. (The practice session was structured exactly like the experimental sessions so as not to confuse the subject.) Each session had four parts: pretest audiogram, fitting and fit-testing of the earmuff, signal detection task and post-test audiogram. Pre- and post-test audiograms (at the frequencies of 500, 2000, 3000 and 4000Hz in each ear) were performed, not only to ensure that the subject's hearing had not changed drastically since the previous session (as a result of tinnitus, head cold, temporary threshold shift (TTS) and so on), but also to determine if the subject experienced TTS as a result of his or her participation in the experiment. (None of the subjects did.)

The adequacy of the earmuff fit across sessions for each subject was evaluated by performing a physical noise reduction (NR) measurement at 1000Hz before the start of each session. For these tests, miniature microphones (Knowles Model BT-1759) were placed in the concha of each ear and on the exterior of each earcup and connected to the L•D 3100 RTA. If the obtained NR measures were such that the highest measured value was no more than 20 per cent greater than the minimum measures obtained, the fit was considered acceptable. (With earmuff attenuation typically ranging from 30 to 35dB at 1000 Hz, allowing 6–7dB of variability in the fit of the earmuff was considered to be a practical compromise between the need for a consistent fit and the reality of achieving that consistency.) If the measures were not in this range (either too high or too low), the earmuff was refitted and the NR measurement repeated until an acceptable fit was achieved.

The experimental task itself consisted of six blocks of 84 individual trials (a total of 504 trials) during which a brief period of noise was presented to the subject. Exactly half of the trials in each block contained a signal (consisting of two 'on' cycles of the reverse alarm) in addition to the noise. The subject's task was to indicate whether or not a signal was heard and how sure he or she was of his or her response. Responses were made by pressing one of six switches on the keyboard located in front of them. Each switch represented a response, ranging from 'Definitely Did Not Hear Signal' to 'Definitely Heard Signal'.

Each of the 12 experimental sessions was devoted to a single, unique combination of the noise level and signal-to-noise ratio variables. In any given session, the noise level and signal-to-noise ratio remained constant (that is, neither the noise level nor the signal-to-noise ratio was changed across the six blocks of trials within a session). The sessions were broken up into six blocks in an effort to reduce the monotony of the task and to allow the subject to take a break if necessary.

Each presentation was preceded by a warning message appearing on the monitor to alert the subject to the fact that a trial was imminent. Immediately following the warning, the test stimulus was presented. The length of this presentation interval was approximately 2s. Coincident with the initiation of the noise, a large question mark appeared on the monitor to indicate that the subject could respond. Two seconds after the noise was terminated, the question mark disappeared (indicating the end of the response completion interval) and one of two feedback messages appeared to indicate whether there had or had not been a signal present in the previous trial. These messages were displayed for approximately 1s. The feedback messages were then removed and the warning message appeared once again. This cycle was repeated until all 84 trials had been completed. The subject was given an opportunity to rest between each block of trials. The structure of the trial blocks (pre-session signal preview, number of blocks, trials per block, total number of trials, warning and feedback) were based on the recommendations of Green and Swets (1988).

Results

At the end of each experimental session, the raw data consisted of a record of what response choice was made by the subject for each of the 504 trials and whether or not a signal had been presented during each trial. Before any analysis could be performed, it was necessary to generate ROC curves for each subject and calculate the area beneath each curve, P(A), the dependent measure used in the subsequent analyses. A detailed discussion and example calculations of these procedures are found elsewhere (Robinson and Casali, 1995) and are not presented here.

Dependent measures for each subject in each experimental condition were calculated using the data from all six blocks of trials as well as the data from just the last five blocks of trials. This was done since discarding the first block of data is common in TSD research (Robinson and Watson, 1972). The data thus obtained were subjected to a paired *t*-test to determine if use of the data obtained in the first block of trials made any difference in the dependent measure. Results indicate that there was not a significant difference between the P(A) measures when the data from the first block of trials were included or excluded (t_{141} = –0.825, p = 0.4110). For this reason, the remaining analyses were conducted using data from all six blocks of trials.

Analysis of Variance

Significant interactions and main effects included signal-to-noise ratio by hearing level (S/N x HL; F = 5.73, p = 0.0049), hearing level (HL; F = 18.81, p = 0.0006) and signal-to-noise ratio (S/N; F = 49.44, p = 0.0001). Since the S/N x HL interaction involves both main effect terms, confounding the main effects, only the interaction is discussed here. (Whenever a main effect is involved in an interaction so that the differences observed in the dependent measure across the main effect variable differ across the other independent variable(s) involved in the interaction, the main effect is said to be confounded and discussion of the main effect outside the context of the interaction is inappropriate.)

Post hoc tests of the S/N x HL interaction were conducted using simple effect *F*-tests followed up by Student–Newman–Keuls tests to determine the locus of each of the simple main effects. These post hoc tests revealed that the performance of the three groups of subjects differed significantly at the three lowest signal-to-noise ratios, but not at the 0dB signal-to-noise ratio. The pattern of significance (at $p \leq$ 0.05) of the simple effects of HL at each level of S/N is shown in Figure 3.3. At signal-to-noise ratios of –8dB and –16dB, individuals with the greatest hearing loss (group 3) performed significantly worse than individuals in the other two groups. At a signal-to-noise ratio of –24dB, individuals in both groups 2 and 3 performed significantly worse than normal listeners, but their performance did not differ significantly from one another.

Though not explicitly labelled in Figure 3.3, the simple effects of S/N for each level of HL were also examined. In this case, it was determined that normal listeners and listeners with only a slight hearing loss exhibited significantly degraded performance only at a signal-to-noise ratio of –24dB. Listeners with mild-to-marked hearing loss showed significant decrements in performance at each level of signal-to-noise ratio, with the exception that their performance at S/N = –16dB and S/N = –24dB did not differ significantly.

One of the subjects (Subject 13) in the mild-to-marked hearing loss group had thresholds in both ears of approximately 60dBHL. (By contrast, the other subjects in this group had thresholds ranging from 45 to 53dBHL in their better-hearing ear.) Throughout the experiment, this subject's performance indicated that she never really heard the signal. Her data were always clustered close to the negative diagonal of the ROC curve and calculated P(A) values hovered around 0.5,

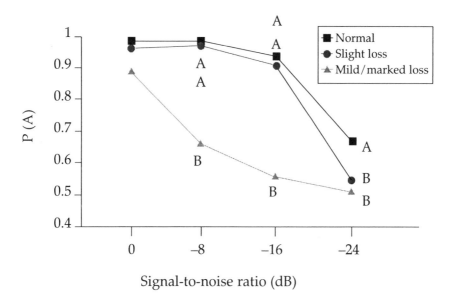

Note: Levels of the HL variable with different letter labels at a given level of S/N are significantly different at $p \leq 0.05$.

Figure 3.3   S/N by HL interaction

indicating chance performance. There was no reason to think that her data were inappropriate for inclusion in the experiment, but there was some concern as to whether or not she belonged in the third group or if she represented yet another, even more hearing-impaired group. For this reason, it was decided to perform the analysis a second time, without this subject's data, to determine how the two analyses differed.

The pattern of significance of the second analysis (excluding Subject 13's data) was identical to that obtained in the first analysis (including Subject 13's data), including the significant differences found in the post hoc analyses, with the exception that the noise level (NL) main effect became significant. Removal of the subject's data caused a slight increase in the mean performance measure on the order of 0.02 to 0.03 at all three noise levels. The data are presented graphically in Figure 3.4.

The Regression Model

In an attempt to gain added insight into the detection problem being investigated, it was decided to develop a regression model in

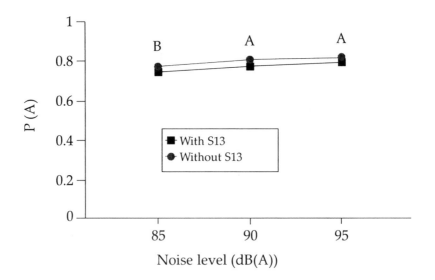

Note: Levels of the NL variable with different letter labels are significantly different at $p \le 0.05$.

Figure 3.4   Noise level main effect, with and without Subject 13's data

addition to performing a traditional analysis of variance. A logistic model which included a natural response frequency term (SAS, 1990), as shown below, was selected to form the basis of this model.

$$P(A) = C + (1 - C)* \left(\frac{e^{f(x)}}{1 + e^{f(x)}} \right) \qquad (3.1)$$

where
$P(A)$ = the area under the ROC curve,
  C = the natural response frequency of the model,
$f(x) = b_0 + b_1x_1 + \cdots + = b_nx_n.$

The $x_1, x_2, \cdots x_n$ terms represent the hearing level (HL), noise level (NL) and signal level (SL) terms as well as the two- and three-way interaction terms. *Signal level* was used as a dependent measure in the model because it was believed to be more intuitive than *signal-to-noise ratio*, which was used in the analysis of variance. A logistic model of this type was chosen for several reasons. First, psychometric functions obtained in auditory experiments often exhibit a sigmoid shape such as that produced by logistic regression models. On inspection, the raw data did appear to exhibit such a shape in several conditions. Second, although threshold theory attributes the sigmoid shape of auditory psychometric functions to the underlying normal distribution of the random variation in the instantaneous threshold (Gescheider, 1985), the logistic distribution produces a very similar curve and is much easier to handle mathematically than is the normal distribution. Finally, inclusion of a natural response frequency would allow the model to approach an asymptote at a value for P(A) of approximately 0.50, which is the minimum possible value of the P(A) term predicted by theory.

Although subjects were screened on the basis of their pure-tone average hearing levels in the range from 500 to 2000Hz, this did not mean that this range would be optimal for the purposes of the regression analysis. It was therefore decided to investigate several methods for quantifying hearing level in the analysis and determine which method provided the most appropriate model. Six schemes for quantifying hearing level were investigated: (1) binaural PTA hearing level over the frequency range from 500 to 2000Hz, (2) minimum monaural (left or right) PTA hearing level over the frequency range from 500 to 2000Hz, (3) binaural PTA hearing level over the frequency range from 500 to 4000Hz, (4) minimum monaural PTA hearing level over the frequency range from 500 to 4000Hz, (5) binaural PTA hearing level over the frequency range from 1000 to 2000Hz, and (6) minimum monaural PTA hearing level over the frequency range from 1000 to 2000Hz. Binaural PTA hearing levels were calculated by multiplying the PTA in the better ear by five, adding this to the PTA of the poorer ear and dividing the result by six. This method was used since it emphasises that both ears are important in an auditory detection task, but that the better hearing ear is *more* important when a large binaural difference exists. This is similar to the methodology used for determining binaural hearing impairment discussed by Miller and Wilber (1991). Thresholds at 1500Hz were included since the alarm used in the experiment had much of its energy in this region.

The first step used in the regression analysis was to fit the full model (which included all of the main effect and interaction terms as well as the intercept term) for each of the six hearing level schemes mentioned above and examine the mean square error (MSE). The MSEs for the models using the PTA over the frequency range from 500 to 4000Hz were nearly double the MSEs for the other four models. It was therefore decided that these two models would be dropped and only the four models with the lowest MSEs would be pursued further.

A backward elimination procedure (Neter *et al.*, 1989) was used to refine the models. This involved the systematic elimination of main effect and interaction terms from the models and examination of the subsequent MSE terms. Elimination of unnecessary terms makes the resulting models easier to understand and explain, simpler to implement, and it reduces the variance of the predictions (Montgomery and Peck, 1982). The natural response term, C, was not considered to be a candidate for exclusion. To do so would have forced the models to approach an asymptote at zero, rather than at a level near P(A) = 0.5.

It was ultimately determined that three terms could be dropped from the model without excessive inflation of the MSE while still being able to achieve a reasonable fit to the experimental data. Fortunately the models for each of the four schemes for quantifying hearing level included the same terms, strengthening the argument that the procedures used to refine the models were correct. The terms included in the final models were hearing level (HL), noise level (NL), signal level (SL), the two-way HL x SL interaction and the three-way HL x NL x SL interaction. The terms dropped from the models included the intercept term as well as the two two-way HL x NL and NL x SL interactions. The equation of the final model is given below.

$$P(A) = C + (1 - C) * \left(\frac{e^{(b_1{}^*HL + b_2{}^*NL + b_3{}^*SL + b_4{}^*HL^*SL + b_5{}^*HL^*NL^*SL)}}{1 + e^{(b_1{}^*HL + b_2{}^*NL + b_3{}^*SL + b_4{}^*HL^*SL + b_5{}^*HL^*NL^*SL)}} \right) \quad (3.2)$$

The differences in magnitude of the MSEs for the four candidate models were quite small, with the MSE for the model based on the binaural PTA hearing level from 1000 to 2000Hz being smallest. This result was not surprising since the alarm had most of its energy in the 1000 to 2000Hz range and, because the screening criteria were so strict, none of the experimental subjects exhibited large binaural hearing level differences. Parameter estimates and the MSE for this condition appear in Table 3.2. Parameter estimates for the other three models can be found in Robinson (1993).

Discussion

Limitations and Constraints on Generalisation

Before the significance of the results presented in the previous section can be discussed, it is necessary to point out the limitations of the research effort and how they may affect interpretation of the results. Most obvious is the fact that, out of the thousands of potential subjects contacted through classified advertisements in three local newspapers, flyers and word of mouth, only two subjects with a slight hearing loss meeting the requirements for the middle group could be recruited for the experiment. The effect this had on the analysis was to reduce the overall power $(1 - \text{ß})$ of the ANOVA and thus increase the potential for making a type II error (ß – failing to reject the null hypothesis when in fact it is false). Also, the fact that both of these subjects' hearing levels placed them at the lower (better hearing) end of the range defining the middle group may have resulted in the mean performance for the

Table 3.2 Parameter estimates and MSE for the model based on the binaural PTA hearing level from 1000 to 2000Hz

Parameter	Parameter estimate
MSE	0.004521590
C	0.520382132
b_1	1.067938668
b_2	–0.98546813
b_3	1.325599931
b_4	–0.033250017
b_5	0.00016938

middle group being higher than would have been the case if individuals with more varied hearing levels had been found.

It would also be desirable, in a follow-on study, to add to the overall number of subjects represented in the data. If more subjects had been available, the power $(1 - ß)$ of the subsequent statistical tests would have been greater. However including more subjects in the experiment was not possible because all recruitment avenues were exhausted within the time and budget constraints. *All* of the hearing-impaired subjects meeting the a priori qualifications for the experiment who were willing to participate in the study were used as subjects; there were no *extra* subjects who met the hearing loss requirements.

In a similar vein, owing to the small number of qualified, hearing-impaired subjects, it was not possible to select subjects on the basis of the origin or cause of the loss. Therefore the results could possibly be confounded by etiology. For example, of the two subjects in the middle group, one exhibited a fairly flat audiogram indicative of conductive hearing loss. The other subject's audiogram, which showed a progressive increase in thresholds at higher frequencies, suggested that his hearing loss was primarily due to presbycusis. Also, all of the subjects in the mild-to-marked group (group 3) experienced tinnitus to some degree, which is quite common as a concomitant symptom with hearing loss. However, this should not be viewed as a serious problem, since all types of hearing loss are represented in an industrial workforce.

Several factors existed which serve to limit the ability to generalise the results of the research described herein. The first of these is the choice of alarm. As mentioned earlier, the specific alarm used in the study was chosen because it had been identified as an extremely common alarm in many industrial facilities and construction sites. The results of the experiment should apply equally well to alarms with the same or very similar characteristics (that is, a 1s period, 50 per cent duty cycle, energy contained primarily in the 1000 to 1500/2000Hz range). However, the results cannot be generalised to different alarms such as a siren or bell, or even a horn, with characteristics which differ considerably from those of the alarm used in the study. This would be the case for any *real* alarm tested. But it is felt that the results of this experiment are more generalisable than would have been the case if a pure tone or a third-octave band of noise had been used, both of which have typically been used in other laboratory signal detection studies.

Also worth mentioning is the choice of HPD. An earmuff was chosen for use in the experiment because it was believed an earmuff would have a greater detrimental effect on the detection of an auditory alarm. Although the trends described by the data may apply when other conventional, passive HPDs are used, the specific results should not be applied to other HPDs, or even to other earmuffs which exhibit attenuation characteristics substantially different from those of the earmuff used in the study.

Analysis of Variance

As can be seen in Figure 3.3, the performance of the subjects in each hearing level group decreases as signal-to-noise ratio decreases (the simple main effect of S/N for each HL group). The point at which performance drops below what would be an acceptable level in an industrial situation (in which the alarm was actually being listened for) differs across hearing groups. For the better-hearing subjects (groups 1 and 2), performance does not drop off drastically until the signal-to-noise ratio drops to –24dB. For the more hearing-impaired subjects (group 3), performance is unsatisfactory at all signal-to-noise ratios other than 0dB. (Although the figure shows the mean performance of the subjects in group 3 (mild-to-marked loss) at the 0dB signal-to-noise ratio to be at a level of P(A) = 0.89, when Subject 13's data are eliminated, the performance of this group improves to P(A) = 0.97. Although the mean performance level of this group also improves slightly at the three lower signal-to-noise ratios when Subject 13's data are eliminated, the performance remains below P(A) = 0.75.) The nature of this simple main effect of S/N for the subjects exhibiting only a slight hearing loss (group 2) might have been slightly different had more subjects fitting into this category been found and had their hearing levels varied over the entire range allowed for this group.

   It is important to note that even subjects with a fairly severe hearing loss (having hearing levels of 45–50dBHL in the frequency range of the signal) are still capable of detecting the signal used in the study when presented at a fairly low signal-to-noise ratio (0dB). It would appear that at this signal-to-noise ratio, across all noise level conditions, the hearing level at which a performance decrement begins to become apparent may fall in the range from approximately 50 to 60dBHL. This statement is based on the finding that the subjects with hearing levels ranging from 45 to 53dBHL had no trouble hearing the alarm in the 0dB signal-to-noise ratio conditions while the subject with a hearing level of approximately 60dBHL could not hear the alarm in any of the experimental conditions. The threshold (assuming P(A) = 0.75 to represent 'threshold') for this signal in the conditions examined in the experiment appears to be at a signal-to-noise ratio of about –8dB. Threshold for the normal-hearing subjects, on the other hand, appears to be at a signal-to-noise ratio between –16 and –24dB.

As expected, there were significant differences found in the performance of the three hearing level groups at three of the four levels of signal-to-noise ratio (–8, –16 and –24dB). However, the nature of the differences involving the group of subjects showing only a slight hearing loss (the middle group) might have been different if more subjects belonging to this group had been present in the study and if these subjects had better represented the range of hearing allowed for this group. It is believed that, as hearing level increases (hearing becomes worse), performance would drop off first at the lowest signal-to-noise ratios (as evidenced by the significant difference between groups 1 and 2 at the –24dB signal-to-noise ratio) and then progress to higher signal-to-noise ratios with the degree of the performance decrement dependent on hearing level at any given signal-to-noise ratio. However, the study described herein can only allude to this effect owing to the small number of subjects and the gap in the data caused by the missing subjects. If more subjects had been available, more and narrower hearing level categories utilised, and more levels of signal-to-noise ratio used, the data would have been more persuasive in this regard.

The noise level effect (illustrated in Figure 3.4), although not significant in the original analysis, was significant when the data were analysed after eliminating the data obtained from Subject 13. The effect was such that the performance of the subjects in the 85dB(A) noise level was slightly poorer than their performance at the other noise levels (90 and 95dB(A)). This effect may have been due to the subjects in the most hearing-impaired group performing slightly worse at the 0dB signal-to-noise ratio in 85dB(A) noise than at the 0dB signal-to-noise ratio in the other noise levels. If this is the case, it is likely caused by the fact that the signal in that condition was at such a low level that the earmuff attenuated it to an extent that approached threshold for the listeners in this group.

This leads to the suspicion that there may be additional factors important in determining whether or not an individual can hear a signal in noise while wearing an HPD, perhaps in the form of the three-way NL x S/N x HL interaction. As mentioned several times previously, it may be that with more subjects (resulting in a more powerful test) this interaction will indeed be shown to be significant.

The Regression Model

Although the dependent measure, P(A), used in the experiment may be interpreted as an indication of the *accuracy* with which an individual would be able to correctly discern a signal in a noisy environment, it is prudent to simply consider a value of P(A) = 0.75

as a 'threshold' value since this value represents a level of performance halfway between chance and perfect performance. In this way, the model may be considered similar to what might have been developed if a more traditional psychophysical procedure had been used, such as the method of constant stimuli. Only the model quantifying hearing level based on the binaural PTA from 1000 to 2000Hz will be discussed. The reasons for this model being preferred over any others are that it emphasises the point that *both* ears are important in such a detection task, and that detection performance depends primarily on the hearing level in the frequency range containing most of the alarm's energy.

As presented earlier, the regression model is of the form:

$$P(A) = C + (1 - C)* \left(\frac{e^{(b_1*HL+b_2*NL+b_3*SL+b_4*HL*SL+b_5*HL*NL*SL)}}{1 + e^{(b_1*HL+b_2*NL+b_3*SL+b_4*HL*SL+b_5*HL*NL*SL)}} \right) \quad (3.2)$$

where
C = the natural response frequency of the model,
HL = the PTA hearing level,
SL = the A-weighted SPL (sound pressure level) of the signal,
NL = the A-weighted SPL of the noise,
$HL*SL$ = the product of the HL and SL terms,
$HL*NL*SL$ = the product of HL, NL and SL terms, and
b_n = estimates of the regression parameters given in Table 3.2.

With the model in this form, it is possible to predict if an individual would be capable of hearing an alarm under known conditions simply by substituting the hearing level, signal level and noise level terms into the model. For example, given a binaural PTA hearing level of 47dBHL (in the frequency range of 1000 to 2000Hz), 90dB(A) noise and a signal level of 82dB(A), the P(A) predicted by the model would be 0.85. This value is above that considered to be a 'threshold' value ($P(A) = 0.75$), but not substantially greater. It would therefore be assumed that the individual was capable of hearing the alarm, but not with absolute certainty.

It is also possible to rearrange the terms of the model and estimate the masked 'threshold' of the signal for an individual with a known hearing level in a given noise. The equation then becomes:

$$SL = \frac{\ln \left(\dfrac{P(A) - C}{1 - P(A)} \right) - b_1*HL - b_2*NL}{b_3 + b_4*HL + b_5*HL*NL} \quad (3.3)$$

where the terms are as specified earlier.

Using this equation and assuming P(A) = 0.75, NL = 90 and HL = 47, the predicted masked threshold for the signal is 80dB(A). As expected, the masked threshold predicted using this equation is slightly less (2dB) than the signal level used in the previous calculation of detectability. To ensure that the signal is audible under most circumstances, including periods of inattention, a safety factor of 10 to 15dB would be added to this masked threshold estimate (Sorkin, 1987; Wilkins and Martin, 1978).

In a similar fashion, the equation can be rearranged and solved for either the HL or the NL terms. Doing so would allow the end user to determine what minimum hearing level should be required of individuals expected to be exposed to known noise conditions, or to determine the level to which the ambient noise must be reduced so that existing alarms or warnings can be heard by all of the workers in the area.

An original goal of the research effort described herein was to develop an empirical model which could be used to predict if an individual with a known hearing loss would be able to hear an auditory alarm in noise while wearing an HPD. This model would require as inputs only data that were easily obtained by individuals responsible for administering a hearing conservation programme (that is, the broadband, A-weighted noise and signal levels and the PTA hearing level of the listener). This is in contrast to existing models and methods for predicting masked thresholds which all require at least a knowledge of the noise and signal spectra and the spectral attenuation characteristics of the HPD. Furthermore, few existing models take the listeners' hearing levels into account. Such a model was also seen as having some utility in determining what the level of an alarm must be if it were to be heard in a known noise by individuals suffering from a specified hearing loss while wearing HPDs. However, owing to the limitations and constraints discussed earlier, this may have been an overly ambitious goal.

Although the model presented here may not be suitable for everyday application in the workplace, it can be compared to other models and methods currently available for calculating masked thresholds. It is in this context that the remaining discussion of the model will be framed. The existing masked threshold calculation method to which the model will be compared is that contained in ISO 7731-1986 (ISO, 1986).

ISO method for predicting a masked occluded threshold   International Standard 7731-1986(E), *Danger Signals for Work Places – Auditory Danger Signals* (ISO, 1986) presents a method for calculating the *unoccluded* masked threshold for a signal in noise based on either octave or one-third octave noise data. The steps are as follows.

Step 1 Starting at the lowest full or one-third octave band level available, the masked threshold (L_{T1}) for a signal in that band is:

$$L_{T1} = L_{N1},$$

where L_{N1} is the noise level measured in the full or one-third octave band in question.

Step 2 For each successive full or one-third octave band filter n, the masked threshold (L_{Tn}) is the noise level in that band (L_{Nn}) or the masked threshold in the preceding band (L_{Tn-1}) less a constant, whichever is *greater*:

$$L_{Tn} = \max(L_{Nn}; L_{Tn-1} - C),$$

where $C = 7.5$dB for octave band data or 2.5dB for one-third octave band data.

If one or more components of a given signal meet or exceed the threshold estimates calculated using these procedures, the signal should be audible. However, to ensure audibility, it is advisable that the signal level be 10 to 15dB above its threshold level. This procedure is unique in that it takes upward spread of masking into account by comparing the level in the band being considered to the level in the preceding band.

To estimate an *occluded* masked threshold for a signal, it is necessary first to subtract the spectral attenuation of the HPD from the noise spectrum. Doing so gives an approximation of the noise spectrum underneath the HPD. Using this noise spectrum and calculating threshold levels as outlined above provides an estimate of the masked threshold of the signal *under* the HPD. In other words, it provides an estimate of the masked threshold of the signal if both the noise and signal were reduced by an amount equal to the HPD's attenuation. Finally, to determine the actual signal levels necessary in the workplace, the spectral attenuation of the HPD must be added to the calculated signal threshold levels. As before, this is an estimate of the signal's threshold level and 10 to 15dB should be added to the results to ensure audibility.

Accounting for hearing loss on the part of the listener is more difficult. One option is to assume that, if the signal's threshold levels under the HPD are above the listener's hearing threshold, the listener should be able to hear the signal. To make this comparison, it is necessary to convert the listener's threshold levels expressed in dB hearing level to dB sound pressure level.

An example of how the ISO procedures can be used to calculate the masked occluded threshold for the reverse alarm described earlier appears in Table 3.3. For the calculations, a uniform pink noise with a broadband level of 92dB(A) (94.9dB) is assumed. In most industrial settings, HPDs would be worn in noise of this level. Furthermore it is assumed that a Bilsom Viking earmuff is being used and that its attenuation characteristics are as shown in Table 3.1.

If one or more of the one-third octave-band components of the reverse alarm exceeds the level specified in the right-hand column of Table 3.3 (the 'environmental threshold'), it should just be audible (that is, at threshold) to an individual with normal hearing. As such, if the overall signal level were such that the most prominent one-third octave-band component (that component centred at 1250Hz – see Figure 3.1) was equal to the 'environmental threshold' depicted in Table 3.3 (88.4dB), the one-third octave-band levels of each of the four most prominent components (1000, 1250, 2000 and 2500Hz) of the reverse alarm would be 85.4, 88.4, 76.4, and 78.4dB, respectively. The broadband level of the alarm would be 90.6dB (91.1dB(A)). Although the above example utilised one-third octave-band noise and attenuation data, masked occluded thresholds may still be estimated if only octave-band noise or attenuation data are available.

It must be noted that caution should be exercised when selecting HPD attenuation data for use in predicting or estimating masked occluded thresholds using a method such as that outlined above. It has been shown that the manufacturer-supplied HPD attenuation data (obtained under ideal laboratory conditions) can overestimate the actual in-service attenuation of the device (Park and Casali, 1991). Even the one-third octave-band data used in the example do so since they were obtained under carefully controlled laboratory conditions using an experimenter-fit protocol intended to minimise variability in the fit of the HPD, which could adversely affect the results of the experiment. The use of such liberal attenuation estimates could result in predicted masked thresholds that are actually too low. Since realistic and reliable HPD attenuation data are not readily available, the only solution may be to use the attenuation data that are available and then add a safety factor to the resulting threshold estimate to account for the better-than-expected attenuation.

Comparing the results   Using the form of the regression model shown in equation (3.3), for a noise level of 92 dB(A), an individual with normal hearing (assume a binaural PTA hearing level of 15dBHL in the 1000 to 2000Hz range, a level well within the range generally accepted to represent 'normal' hearing (Miller and Wilber, 1991)), the occluded masked threshold of the reverse alarm would be 72dB(A). This estimate is almost 19dB less than that arrived at using the ISO procedures presented above. There are several potential explanations

Table 3.3   The masked occluded threshold for the reverse alarm

One-third octave band freq. (Hz)	Pink Noise One-third octave band level (dB)	One third octave band noise levels under HPD (dB)*	masked threshold under HPD (dB)	Environmental signal levels required (dB)**
25	80.1	—		
31.5	80.1	—		
40	80.1	—		
50	80.1	—		
63	80.1	—		
80	80.1	—		
100	80.1	70.9	70.9	—
125	80.1	70.1	70.1	—
160	80.1	68.1	68.1	—
200	80.1	63.7	65.6	—
250	80.1	59.3	63.1	—
315	80.1	57.1	60.6	—
400	80.1	52.7	58.1	—
500	80.1	48.1	55.6	—
630	80.1	43.7	53.1	—
800	80.1	38.8	50.6	—
1 000	**80.1**	**36.0**	**48.1**	**92.2**
1 250	**80.1**	**37.3**	**45.6**	**88.4**
1 600	80.1	37.8	43.1	—
2 000	**80.1**	**39.1**	**40.6**	**81.6**
2 500	**80.1**	**41.3**	**41.3**	**80.1**
3 150	80.1	36.9	38.8	—
4 000	80.1	31.6	36.3	—
5 000	80.1	33.6	33.8	—
6 300	80.1	36.9	36.9	—
8 000	80.1	36.7	36.7	—
10 000	80.1	—		
12 500	80.1	—		
16 000	80.1	—		
20 000	80.1	—		

Notes

* Calculated using the 1/3 OB attenuation values appearing in Table 3.1.
** To avoid confusion, only the 'threshold' levels for the four most prominent 1/3 OB components (shown in **bold** type) of the reverse alarm are presented.

Source: Procedures outlined in ISO 7731-1986 (ISO, 1986).

for such a discrepancy. Although not stated explicitly, the method presented in ISO 7731-1986 (ISO, 1986) for calculating masked thresholds is most likely based on critical band theory to some degree. This theory, originally developed by Fletcher (1940), has been extended and modified extensively by subsequent research (Scharf, 1970). In this process, many different experimental paradigms have been used and various stimuli employed as the test signals and maskers including pulsed and continuous pure tones, pulsed and continuous bands of noise, as well as broadband white and pink noise, to name a few. The resulting theory cannot be expected to be absolutely accurate for all real-world situations. These factors could be responsible for at least some of the differences between the two predictions.

Furthermore, much of the research upon which critical band theory is based was conducted using relatively low stimulus levels while the test subjects were unoccluded (that is, they were not wearing an HPD). The experiment described herein was conducted using high stimulus levels while the subjects were wearing HPDs. It is possible that the enhanced bone conduction due to the use of the HPDs interacted with the high stimulus levels to reduce the masked thresholds under the hearing protectors. Finally, in any regression problem, the model is intended to represent a 'best fit' and will not pass through all of the data. Some of the data will fall above the line or surface defined by the model while other data will fall below this line or surface. Quite often a predictive model is based on a 90 or 95 per cent confidence level, or the 90 or 95 per cent confidence interval (or the equation for calculating the interval) is presented in addition to the predicted value or model itself. However, because of the manner in which the data were obtained and the non-linear form of the model used, it is not possible to calculate or present such a confidence interval. If it were possible to do so, the thresholds predicted by the model would be somewhat higher and therefore closer to the threshold predicted using the ISO standard method.

In any case, the differences between the two predictions do not render the results of the experiment or the subsequent regression model worthless. This experiment does provide some empirical evidence that occluded masked threshold predictions obtained using the ISO standard procedures are somewhat conservative, at least for the conditions used in the experiment, and tend to overestimate the actual occluded masked threshold.

Conclusions

Perhaps the most important conclusion reached based on the results of the research effort described herein is the fact that individuals with a substantial hearing loss (on the order of 45 to 50dBHL) are capable of hearing a reverse alarm presented at a reasonably low signal-to-noise ratio (0dB) while wearing a high-attenuation earmuff. However, when considering individuals with greater hearing loss or noise levels other than those investigated in the experiment, the picture is less clear. The marginal noise level effect seems to indicate that, at noise levels less than 85dB(A), individuals with a mild-to-marked hearing loss may begin to experience difficulty in detecting such an alarm at such a low signal-to-noise ratio. Also detectability of such an alarm in noise while wearing an HPD depends, at the very least, on the signal-to-noise ratio and the hearing level of the individual and probably also on the level of the masking noise. At reasonable signal-to-noise ratios (0dB), the hearing level at which a performance decrement begins to become apparent seems to be in the range between approximately 50 and 60dBHL. It would appear that, if the recommendation of paragraph 5.2.1 of ISO 7731-1986 (ISO 1986) (the signal should be audible if the A-weighted signal level exceeds the background noise level by 15dB) is followed, individuals with hearing losses on the order of 45–50dB will indeed be able to hear alarms or warnings such as the one used in this experiment even when wearing an earmuff.

When comparing the results of the experiment reported here with masked occluded thresholds calculated using the procedures specified in ISO 7731-1986 (ISO, 1986), it would appear that the thresholds so obtained are conservative and would likely be more than adequate estimates for a large portion of the population. This can only be said with any certainty, however, for similar situations where similar signals are used, where the noise is relatively flat and falls within the range of noise levels used in the experiment, and an earmuff with similar attenuation characteristics is used. If conditions differ substantially from those used in the experiment, this conclusion may not be justified.

Recommendations

Although the experiment described here was successful, the scope of the research was limited in that only one alarm, one noise and one HPD were used and that it was not possible to screen and diagnose subjects according to etiology of hearing loss. Additional experiments should be conducted which investigate the issue of audibility of auditory alarms in noise, but they should include more subjects and narrower hearing level categories (on the order of 10 to 15dB). Additional independent variables should also be investigated, including etiology, protector type (plug versus muff and flat-attenuation devices) with attenuation as a variable, alarm type (that is, bell, siren, horn) and spectra, and attentional demand. Although it would not be possible to include all of the above dimensions in a single study, it may be possible to use sequential experimentation techniques and answer the research questions as part of a multi-year effort. Since criteria would not be investigated, the two-interval forced-choice procedure of signal detection theory would probably be the best and most efficient method of data collection.

Furthermore, the example calculations presented in the discussion of the regression model serve to illustrate just how cumbersome existing methods for estimating a masked occluded threshold can be and how much information such methods require. What is needed is a simplified method based on broadband SPL measurements and simple hearing level classifications. (For instance, PTA thresholds or hearing levels in the better ear instead of estimates of auditory filter width used in some models.) Such a simplified model would be more likely to find general acceptance in industry than would a model requiring information not generally available to the industrial hygienist or hearing conservationist, or that would require additional hearing tests or sound level measurements, necessitating additional expenditures of both time and money.

Acknowledgments

This research effort was sponsored by the Corporate Health Department of the Aluminum Company of America (ALCOA), Pittsburgh, Pennsylvania. Christine Dixon-Ernst is thanked for her input and for serving as ALCOA technical monitor.

References

Abel, S.M., Kunov, H., Pichora-Fuller, M.K. and Alberti, P.W. (1985) 'Signal detection in industrial noise: Effects of noise exposure history, hearing loss and the use of ear protection', *Scandinavian Audiology*, **14**, 161–73.

Casali, J.G and Robinson, G.S. (1990) *A reverberant computer-controlled facility for hearing protection research and attenuation testing: Verification re ANSI S3.19-1974* (ISE Technical Report No. 9007, Audio Lab No. 12/1/90/5-HP). Blacksburg, VA: Auditory Systems Laboratory, Department of Industrial and Systems Engineering, Virginia Polytechnic Institute and State University.

Casali, J.G., Mauney, D.W. and Burks, J.A. (1995) 'Physical vs. psychophysical measurement of hearing protector attenuation – a.k.a. MIRE vs. REAT', *Sound and Vibration*, **29**(7), 20–27.

Coleman, G.J., Graves, R.J., Collier, S.G., Golding, D., Nicholl, A.G.McK., Simpson, G.C., Sweetland, K.F. and Talbot, C.F. (1984) *Communications in Noisy Environments* (Report Number TM/84/1), Edinburgh: Institute of Occupational Medicine, Ergonomics Branch.

EPA (1981) *Noise in America: The extent of the noise problem* (EPA Report 550/9-81-101), Washington, D.C.: Environmental Protection Agency.

Fidell, S. (1978) 'Effectiveness of audible warning signals for emergency vehicles', *Human Factors*, **20**, 19–26.

Fletcher, H. (1940) 'Auditory patterns', *Reviews of Modern Physics*, **12**, 47–65.

Forshaw, S.E. (1977) *Listening for machinery malfunctions in noise while wearing ear muffs* (DCIEM Technical Report No. 77x43), Ontario: Department of National Defense, Defense and Civil Institute of Environmental Medicine, Behavioral Division.

Gescheider, G.A. (1985) *Psychophysics – Method, Theory and Application*, 2nd edn, Hillsdale, NJ: Lawrence Erlbaum Associates.

Green, D.M. and Swets, J.A. (1988) *Signal Detection Theory and Psychophysics*, Los Altos, CA: Peninsula.

ISO (1986) *Danger Signals for Work Places – Auditory danger signals* (ISO 7731-1986(E)), Geneva: International Organization for Standardization.

Laroche, C., Tran Quoc, H., Hétu, R. and McDuff, S. (1991) '"Detectsound": A computerized model for predicting the detectability of warning signals in noisy environments', *Applied Acoustics*, **33**(3), 193–214.

Lazarus, H. (1980) 'The effects of hearing protectors on the perception of acoustic signals', *Zentralblatt für Arbeitsmedizin*, **30**, 204–12.

Miller, M.H. and Wilber, L.A. (1991) 'Hearing evaluation', in C.M. Harris (ed.), *Handbook of Acoustical Measurement and Noise Control*, 3rd edn. New York: McGraw-Hill.

Montgomery, D.C. and Peck, E.A. (1982) *Introduction to Linear Regression Analysis*, New York: Wiley.

Neter, J., Wasserman, W. and Kutner, M.H. (1989) *Applied Linear Regression Models*, (2nd edn), Homewood Il: Irwin.

NIH (1990) *Noise and Hearing Loss* (NIH consensus development conference statement 22–24 Jan, 8(1)), Bethesda, MD: National Institutes of Health.

Park, M.Y. and Casali, J.G. (1991) 'A controlled investigation of in-field attenuation performance of selected insert, earmuff and canal cap hearing protectors', *Human Factors*, **33**(6), 693–714.

Patterson, R.D. (1974) 'Auditory filter shape', *Journal of the Acoustical Society of America*, **55**, 802–9.

Patterson, R.D. (1976) 'Auditory filter shapes derived with noise stimuli', *Journal of the Acoustical Society of America*, **59**, 640–54.

Patterson, R.D., Nimmo-Smith, I., Weber, D.L. and Milroy, R. (1982) 'The deterioration of hearing with age: Frequency selectivity, the critical ratio, the audiogram and speech threshold', *Journal of the Acoustical Society of America*, **72**, 1788–1803.

Robinson, D.E. and Watson, C.S. (1972) 'Psychophysical methods in modern psychoacoustics', In J.V. Tobias (ed.), *Foundations of Modern Auditory Theory, Volume II*, New York: Academic Press.

Robinson, G.S. (1993) 'Using signal detection theory to model the detection of warning signals in normal and hearing-impaired listeners while wearing hearing protection', unpublished doctoral dissertation, Virginia Polytechnic Institute and State University, Blacksburg, VA.

Robinson, G.S. and Casali, J.G. (1995) 'Audibility of reverse alarms under hearing protectors for normal and hearing-impaired listeners', *Ergonomics*, **38**(11), 2281–99.

SAE (1978) *Performance, Test and Application Criteria for Electronically Operated Backup Alarm Devices* (ANSI/SAE J994b-1978), Warrendale, PA: Society of Automotive Engineers, Inc.

Scharf, B. (1970) 'Critical bands', in J.V. Tobias (ed.), *Foundations of Modern Auditory Theory, Volume I*, New York: Academic Press.

Sorkin, R.D. (1987) 'Design of auditory and tactile displays', in G. Salvendy (ed.), *Handbook of Human Factors*, New York: Wiley.

Suter, A.H. (1989) 'The effects of hearing protectors on speech communication and the perception of warning signals' (Technical Memorandum 2–89), Aberdeen Proving Ground, MD: US Army Human Engineering Laboratory.

Swets, J.A. (1986) 'Indices of discrimination or diagnostic accuracy: Their ROCs and implied models', *Psychological Bulletin*, **99**, 100–117.

Swets, J.A. (1988) 'Measuring the accuracy of diagnostic systems', *Science*, **240**, 1285–93.

Swets, J.A., Pickett, R.M., Whitehead, S.F., Getty, D.J., Schnur, J.A., Swets, J.B. and Freeman, B.A. (1979) 'Assessment of diagnostic technologies', *Science*, **205**, 753–9.

Wilkins, P.A. (1984) 'A field study to assess the effectiveness of wearing hearing protectors on the perception of warning sounds in an industrial environment', *Applied Acoustics*, **17**, 413–37.

Wilkins, P.A. and Martin, A.M. (1977) 'The effect of hearing protection on the masked thresholds of acoustic warning signals', paper presented at the 9th International Congress on Acoustics, Madrid.

Wilkins, P.A. and Martin, A.M. (1978) *The Effect of Hearing Protectors on the Perception of Warning and Indicator Sounds – A general review*, Technical Report No. 8, Southampton: Institute of Sound and Vibration Research, University of Southampton.

Wilkins, P.A. and Martin, A.M. (1981) 'The effect of hearing protectors on the attention demand of warning sounds', *Scandinavian Audiology*, **10**, 37–43.

Wilkins, P.A. and Martin, A.M. (1982) 'The effects of hearing protection on the perception of warning sounds', in P.W. Alberti (ed.), *Personal Hearing Protection in Industry*, New York: Raven Press.

Wilkins, P.A. and Martin, A.M. (1984) 'Attention demand and recognition in the perception of warning sounds and the effects of wearing hearing protection', *Journal of Sound and Vibration*, **94**, 483–94.

Wilkins, P.A. and Martin, A.M. (1985) 'The role of acoustical characteristics in the perception of warning sounds and the effects of wearing hearing protection', *Journal of Sound and Vibration*, **100**, 181–90.

Wilkins, P.A. and Martin, A.M. (1987) 'Hearing protection and warning sounds in industry – A review', *Applied Acoustics*, **21**, 267–93.

4 Extending the Domain of Auditory Warning Sounds: Creative Use of High Frequencies and Temporal Asymmetry

R.D. PATTERSON, *Centre for the Neural Basics of Hearing, Physiology Department, University of Cambridge* and **A.J. DATTA,** *MRC Applied Psychology Unit, Cambridge*

Introduction

Several years ago, at the request of RAE FS9 (now DERA Ae FS9), the Applied Psychology Unit (APU) prepared a set of 12 auditory warning sounds for use in military helicopters to signal potential problems in flight systems. The sounds were prepared in accordance with guidelines that are now summarised in Patterson (1990), and they were tested to ensure a lack of confusability by Munger and Rood of DERA Ae FS9. In the original flight systems warnings most of the energy was below 4000Hz. There was a need to increase the frequency range of the auditory warnings to 12000Hz to make them localisable in the advanced audio display unit envisaged by DERA Ae. At the same time, it was important to change our perception of the sounds as little as possible since the existing warnings were already installed in operational aircraft. There was no way of increasing the frequency range of the existing auditory warnings without producing some noticeable change in their sound quality. However we noted (a) that the temporal pattern of a complex sound is a major determinant of its character and (b) that the main contribution of the high-frequency components was to brighten the

73

timbre of the sound. This suggested that a practical solution might be to add high-frequency energy with the same temporal envelope to each of the existing warning sounds. There appeared to be three ways of solving this problem.

First was envelope filling, where the envelope of the existing warning was extracted and applied to a set of high-frequency harmonics. Then the two complex waves were combined in appropriate proportions. The second possibility was Nyquist whistling. Here the existing warning was digitised at a rate just above that required by the main energy band. Then it was replayed without an anti-aliasing filter which introduced high-frequency energy in the form of a reflection of the original spectrum about the half sampling rate. This gave the sounds a distinctive whistling character. The sampling rate has to be tuned to get the right degree of whistling. Finally, there was fine structure doubling; the waveform of the existing warning was segmented into cycles from one zero crossing to the next and each cycle was replaced by two compressed versions of the cycle that fitted the same cycle time. Then the time-compressed sound and the original sound were combined in appropriate proportions to produce the new warning.

This chapter describes the algorithms and tools used to produce a large set of prototype warning sounds based on these generation techniques, and the listening procedures used to evaluate the prototypes in preparation for selecting a final set. The tools are available from the software package associated with the Auditory Image Model (AIM) (Patterson *et al.*, 1995).

Envelope Filling

The envelope filling technique is a form of amplitude modulation in which one signal, the *carrier*, is multiplied by a second signal, the *modulator*. The envelopes of the original auditory warnings are the modulators in this case and the carrier is a set of high-frequency harmonics. The new warning is produced by multiplying the modulator by the carrier. To start with, analogue recordings of the existing warnings were digitised at a sampling rate of 20 000 points per second; the half sampling rate was well above the highest frequency in these warning sounds.

The first task in the envelope filling method, as the name suggests, was to extract the envelope from the original warning sound. The routine in the auditory image software package (Patterson *et al.*, 1995) for generating a spectrogram of a wave was used to extract the envelope. The filter bank and compression were turned off; full wave rectification and low-pass filtering were turned on. The decay time of the low-pass filter was kept short (5ms) to ensure that brief dips in the envelope of the original sound were preserved. A general purpose computer routine was used to generate digital sinewaves and add them to synthesised harmonic sounds. Twenty harmonics of 250Hz, from 6000Hz to 11000Hz, with random phase, were added to produce a high-frequency complex tone (referred to as hfh). The reason for adding in random phase was to avoid creating sound waves with large peak factors. The envelope and the high-frequency harmonics were multiplied to form an amplitude modulated signal. The resultant wave was divided by the maximum value of hfh, to normalise it to the height of the envelope of the original sound. Finally the resultant amplitude modulated waveform was added to the original warning to produce the new prototype warning. Three forms of each prototype warning were produced with the level of the added harmonics having the same, one-half or one-quarter of the energy of the original warning.

Nyquist Whistling

The Nyquist whistling technique is a novel use of the sampling theorem which normally specifies the minimum sampling rate for adequate representation of a continuous signal. The critical sampling rate is twice the rate of the highest frequency component in the signal, and is called the Nyquist frequency. When sounds are recorded at too low a rate, or when they are replayed without an anti-aliasing filter, the original sound is accompanied by a whistling sound at high frequencies. The whistling can be explained by the spectrum of the recorded sound. When a signal is digitised, a copy of the spectrum in the region below the half-sampling frequency appears reflected in the spectrum between the half-sampling frequency and the Nyquist frequency. The effect is known as aliasing and is an unavoidable by-product of digitising a continuous wave. Normally an anti-aliasing filter is used to remove the high-frequency portion of the spectrum.

The Nyquist whistling technique was conducted as follows. The auditory warnings were recorded at sampling rates of 8000, 10 000, 12 000 and 16 000 samples per second. The aim was to locate the sampling rate which would just pass the energy of the auditory warning (without losing too much information), that is, the effective Nyquist frequency. When the warning recorded at 16 000 samples/sec. was replayed, it sounded identical to the original, indicating that most of the energy in the auditory warnings lay below 8kHz. The Datlink interface used to record and play the sounds has a built-in anti-aliasing filter. To neutralise it, each point of the sampled warning was copied n times and played back at n times the recording rate, at which point the whistling effect becomes audible.

Each of the 12 auditory warnings was passed through a routine called '*n*times'. The function '*n*times' took the auditory warning as input and the output was each point of the digitised warning written *n* times in the output stream (where *n* was 1, 2, 3, …). The modified warning was played back at *n* times the recording rate. For example, warnings recorded at a sampling rate of 12kHz were passed to *n*times with argument four, writing each point four times onto the output, and the modified warning was played back at a sampling rate of 48kHz. This technique neutralised the anti-aliasing filter on the Datlink interface and enabled one to hear the Nyquist whistling.

Fine Structure Doubling

Theoretically this method is the most appropriate for adding high-frequency components and producing minimal change in sound quality because it produces a sound like the octave of the original which should blend well with the original in terms of sound quality. However most of the original warnings had irregularly shaped waves within the cycle and rapidly varying cycle times (periods). As a result they strained the algorithm even with a 48 000Hz sampling rate, and the resulting distortion rendered some of the modified warnings unusable.

The aim of the technique was to replace each cycle of the digitised wave with two compressed copies of the cycle. The warning was digitised at a high sampling rate (48 000Hz) and the wave was divided into cycles from one zero crossing to the next. After a cycle had been isolated, every other point was dropped before doubling so that the total number of points per cycle remained constant. If there was a mismatch of one point at the end of the doubled cycle, the distortion was audible. For a cycle with even numbered points, dropping alternate points and doubling the compressed cycle was straightforward. For a cycle with an odd number of points, the last point of the compressed cycle was dropped when it was copied for the second time, to keep the cycle length constant. The original and double waveforms were divided by two so that, when they were added, they stayed within the two byte limit.

Listening Tests

Special purpose, interactive computer programs were written to facilitate listening to the prototype warning sounds in a wide variety of forms in order to compare and contrast them and assist the process of narrowing the choices down to the more suitable forms. The programs are referred to as hhplay, nqplay and dcplay for the high-harmonic, nyquist and double-cycle sounds, respectively.

Hhplay was the listening tool for the *envelope filling* method. The original warning was played first, followed by the high-frequency amplitude modulated sound on its own. Then the original warning and three prototypes were played: (1) the original plus the hfh carrier, (2) the original warning plus the hfh carrier divided by two, and (3) the original warning plus the hfh carrier divided by four. The whole sequence could be repeated *n* times by specifying *n* as the second argument of the shell tool hhplay .

Nqplay was the script file for listening to the signals generated by the 'Nyquist whistling' method. The original warning, recorded at 20 000 samples/sec. was played first, followed by the versions recorded at 12 000 samples/sec., 10 000 samples/sec. and 8000 samples/sec. Then the original warning and three prototypes were played: (1) the warning recorded at 16 000 samples/sec. with each point written three times, (2) the warning recorded at 12 000 samples/sec. with each point written four times, and (3) the warning recorded at 8000 samples/sec. with each point written six times. Finally the original and the warning recorded at 10 000 samples/sec. with each point written twice were played. The number of repetitions could be specified by the listener as the second argument of the tool nqplay.

The final listening tool was dcplay. The original warning and three prototypes were played: (1) the original plus the cycle doubled signal, (2) the original plus the cycle doubled signal divided by two, and (3) the original plus the cycle doubled signal divided by four. The listener could specify the number of repeats using the second argument of dcplay.

Listening Results

Listening tests were performed with the staff from APU and DERA. Judgments of the effectiveness of each method of extending the frequency range were made for each of the existing warnings and the best value for the parameters was noted. The full set of judgments is recorded in Patterson and Datta (1994). This section describes a subset of the results – primarily the best new warnings, but with comments on a few of the worst.

It is very difficult to describe these warnings verbally, perhaps because they were designed to be distinctive. The system for generating them is described in reports by Lower *et al.* (1986) and Patterson *et al.* (1986) which include specific parameter values and a form of notation for describing the warnings. In the remainder of this section, for brevity, the warnings are referred to by their DERA numbers.

Auditory Warning Number 1

The result of envelope filling was not good for this warning. In the original sound, the envelope fluctuated rapidly and the spectrum changed rapidly with the envelope. When the envelope was filled with static high-frequency harmonics, the rapid fluctuations were less audible because there was no concomitant spectral change. The two signals sounded like two separate sources. There was something very shrill about the hfh component in this case. It drew the listener's attention well, but it quickly became aversive.

The Nyquist whistling method did a surprisingly good job of adding high-frequency energy while maintaining the good aspects of the original sound quality. In fact Nyquist whistling worked quite well for most of the warnings. The only decision was to choose the best degree of whistling, that is, the best sampling rate for the extension in the blend. Perceptually, the blend with the 10kHz extension seemed best.

The distortion introduced by the fine structure doubling technique was particularly intrusive with this warning. The added components sounded like a totally different source that had only the modulator in common. In summary, Nyquist whistling with the 10kHz extension seemed best for warning 1.

Auditory Warning Numbers 2, 4, 5, 10

These warnings were similar to the first and produced a similar preference for Nyquist whistling with a 10kHz cutoff.

Auditory Warning Number 3

This warning was well suited to the envelope filling method. The envelope of the original sound had well spaced fluctuations. Hence, when the high-frequency harmonics were added to the warning, the effect was pleasant. Nyquist whistling also worked well, as did the fine structure doubling. The blend with extension 10kHz sounded very good, as did the blend with extension 8kHz. Any one of the three methods would do in this case.

Auditory Warning Number 6

Envelope filling worked well for this warning with hfh at half the level of the primaries. Nyquist whistling was acceptable; the blend with 10kHz extension seemed the best. Fine structure doubling introduced distortion in the extension. When blended to produce new warnings, the added components produced a change in the timbre which was intrusive even when the relative level of the extension was low. The extension also changed the perceived urgency inappropriately. The original warning had well spaced pulses and, as a result, all three methods worked in the sense of producing recognisable blends. Nevertheless envelope filling and Nyquist whistling produced better warnings than cycle doubling.

Auditory Warning Number 7

Both the Nyquist whistling and the fine structure doubling produced a good result. The extensions were not perceived as separate sources in the blend. Rather they brightened the timbre of the warning and gave it extra distinctiveness. Since the cycle doubling method did not produce good blends as often as Nyquist whistling, it would make sense to use cycle doubling in this case to increase the distinctiveness of the warnings within the set.

Auditory Warning Number 8

The original warning sounded like a calliope because it had a breathiness that was reminiscent of air whistling through pipes. The envelope filling extension had no breathiness whatsoever, and in the blends the extension reduced the overall breathiness considerably. As a result, the blends were quite different in character from the original.

The quality of the Nyquist whistling blend with the 10kHz extension was probably better than the original, but all the blends had a 'chirp' that increased their distinctiveness and made all of them more acceptable than the original. The cycle doubling method produced extensions with a strong breathiness and so the prototypes with cycle doubled extensions had more breathiness than the original. The effect was highly satisfactory and, since the cycle doubling method did not produce good blends as often as Nyquist whistling, it made sense to use it in this case as well.

Auditory Warning Number 9

This warning had well spaced pulses, yet the envelope filling method was not very successful for this sound. One could perceive the presence of two separate sources which made it quite different from the original. The shrill character of the extension made it stick out in the blend. New warnings produced by the Nyquist whistling technique had a chirping effect which threatened to dominate the character of the warning. Cycle doubling produced extensions that were not heard as separate sources and which seemed to intensify the natural character of this warning sound. The extensions improved the sharpness of the warning while preserving the fundamental character of the sound. Fine structure doubling produced the best overall results and the different blends all seemed equally acceptable.

Auditory Warning Number 11

The blends produced by envelope filling were all quite bad. Two separate sources were heard because the extension did not have the strong frequency sweep of the original. The timbre of the extension stuck out. It and the original warning were very different. The Nyquist whistling method actually reduced the level of this warning sound. This suggested that the original sound already had high-frequency energy, and that this energy was removed in the recording process when the lower cutoffs were used. If this was the case, there was no need to modify this warning sound.

Fine structure doubling produced the best outcome here. The high-frequency extension produced minimal disruption of sound quality in the blends. The extension increased the sharpness of the sound with the introduction of some noisiness which was generally acceptable.

Auditory Warning Number 12

The envelope filling technique did not work for this warning sound since the original sound had no amplitude modulation. The original sound drew attention through frequency modulation, which could not be captured by this procedure. Hence the extension was clearly distinguishable as a separate source in all three blends. The Nyquist whistling effect was barely audible in blends of this warning sound, indicating that the upper frequencies in this warning lie well below the lowest half-sampling rate (8kHz) and the reflected part of the spectrum is limited to a small region around the Nyquist frequency. In any event, it did not produce a good warning sound.

Fine structure doubling produced the best outcome with this warning; indeed it was probably the best result for this method with any of the 12 warnings. When the high-frequency energy was added, it barely changed the timbre of the warning. The brightness was increased and that was about the only audible change. The technique worked well at all the three levels in the blends.

Auditory Warnings with Temporal Asymmetry

In this phase of the project, the purpose was to develop a new class of sounds to be used as threat warnings. There was also the constraint that they needed to be integrated with the *aircraft systems* warnings and *flight parameter* warnings to form a well structured warning suite. This led to the investigation of a new class of sounds with temporal asymmetry in the envelope. These warnings have a timbre or sound quality different from the existing warnings, so they are distinguishable as a class. Instead of synthesis in the frequency domain, the new sounds are generated in the time domain. Varying individual parameters produces a range of similar sounds with varying degrees of urgency. Unique combinations of envelopes and carriers can be used to produce identifiable, dedicated attensons.

Temporally asymmetric envelopes were applied to a carrier to produce a distinctive new sound. A degree of jitter was also added to the envelope period, or one of the other parameters, to increase distinctiveness and urgency. We report the results of searching the space of parameters and values to find appropriate sounds for aircraft threat warnings. Various complex carriers were used which broadly fell into two categories: (a) complex tones, where harmonics or octaves with random phases were used as constituent sounds, and (b) iterated ripple noise (IRN) which is constructed from a random noise by delaying a copy of the noise, adding it to the original, and iterating or repeating the process a number of times.

Various envelope shapes were considered while investigating asymmetry. Damped and ramped envelopes were chosen as the staring point, since the effect of shape was not dramatic and the damped and ramped envelopes have been studied systematically by Patterson (1994a, 1994b) and Irino and Patterson (1996). The variables which affect damped/ramped sounds are the envelope period, the half-life of the exponential decay, and the amplitude and the floor level where the damped/ramped envelope ends. Randomness (or jitter) was introduced to each of these parameters to enhance distinctiveness. The search space was vast and so we began by developing tools to explore the space systematically, taking one parameter and one carrier at a time, to find a reasonable range.

A Structured Listening Tour through the Space of Asymmetric Sounds

The time-asymmetric envelopes form the basis of the new class of warnings. Four carriers were used to produce four distinct subclasses of sounds. Tools were prepared to present large numbers of these sounds to listeners in a convenient form. The parameters were varied systematically and perceptual descriptions of the sound qualities were noted. The range was surprisingly large; a subset of the tables of results is presented below (Table 4.1). The sounds marked with asterisks are the ones which could form the basis for attensons.

Table 4.1   Sound qualities of prototype samples of octave-spaced sine waves

Octave-spaced harmonic carrier damped sounds

Half-life	Period	Timbre
4ms	45ms	Drumlike organ clicks
4ms	90ms	Pizzicato organ notes
4ms	180ms	Strong, brief organ notes
8ms	45ms	Organ component stronger and rapid flutter
8ms*	90ms	Organ component stronger and flutter
8ms*	180ms	Organ component stronger and slower flutter
16ms	45ms	Tone begins to dominate
16ms*	90ms	Metallic organ taps
16ms	180ms	Metallic organ taps with more damping heard
32ms	45ms	Too tonal, weak clicks
32ms*	90ms	Metallic bell sounds
32ms	180ms	Bell sounds

IRN (lag 16ms) carrier damped sounds

Half-life	Period	Timbre
4ms	45ms	Rapid brief clicks
4ms*	90ms	Brief clicks
4ms	180ms	Brief clicks
8ms	45ms	Snare drum effect
8ms*	90ms	Snare drum effect
8ms	180ms	Snare drum effect
16ms*	45ms	Propeller plane
16ms	90ms	Propeller plane with slow rotation
16ms	180ms	Dull sound

32ms*	45ms	Loud propeller plane
32ms	90ms	Cylinder helicopter
32ms*	180ms	Slow piston-like effect

IRN (lag 16ms) carrier ramped sounds

Half-like	Period	Timbre
4ms	45ms	Noisy flutters/clicks
4ms	90ms	Noisy flutters/clicks
4ms	180ms	Noisy clicks
8ms	45ms	Noisy clicks with little tone
8ms	90ms	Noisy clicks with little tone
8ms	180ms	Noisy clicks with little tone
16ms	45ms	Ship's funnel
16ms*	90ms	Cylinder helicopter
16ms	180ms	Piston with stronger tone
32ms*	45ms	Loud ship's funnel
32ms	90ms	Piston effect pronounced
32ms	180ms	Strong piston with tone

The Effect of Jitter

The effect of jitter seemed to be largely orthogonal to sound quality for these sounds. When jitter was gradually introduced to one of the envelope parameters, the perception of the sound did not change suddenly to that of a new source. As a result jitter was omitted in the initial, parametric, listening tests. Then, once an interesting sound was identified, randomness was introduced in one of the parameters to accentuate the distinctiveness of the sound.

We started our search by creating sounds having a wide range of jitter in exponential steps. The initial range was from 1 to 90 per cent with the steps being 1, 10, 30, 50 and 90. In general:

a The range 1–10 hardly produced any noticeable difference in the sounds.
b Jitter above 80 produced sounds with crackling distortions which were generally disruptive.
c Jitter in the range of 10–30 produced sounds where the effect was gradually noticed. Though psychophysically important, we

thought that this effect was not strong enough to produce distinctiveness in attensons.

d   The ideal range seemed to be 30–70, so exploration of the space was focused on sounds with 30, 50 and 70 per cent jitter.

We listened to a wide range of the aysmmetric sounds with jitter, separately, in all four parameters. It was noted that those sounds which were dull to start with (no asterisks) did not become good attensons by virtue of the addition of jitter. However, for those that had already been chosen as potential attensons, jitter tended to enhance their attention-gaining quality. The effects of jitter in the four parameters are summarised below.

Half-life

Random fluctuations in the half-life seemed to impart rhythmic patterns to the sounds. As the percentage of variation increased, the rhythm became quite strong, but it was never disruptive. For example, the octave-spaced harmonic carrier with a damped envelope (16ms half-life), sounded like metallic organ taps. With a jitter of 50, it was metallic organ taps with a rhythmic pattern – an effect which would probably enhance retention of the sound quality in memory.

Envelope Period

Variations in the envelope period added jumpiness or hesitation to the sound quality. Listening tests did not produce sounds which seemed better than the non-random sounds or half-life jittered sounds. At higher values, crackling distortion was added to the sounds.

Amplitude

As expected, overall amplitude variation randomly changed the loudness of the individual pulses making up the sounds. It was an interesting effect, but probably not useful for attensons.

Floor

Floor variation produced sounds which seemed to have effects of both envelope period jitter and amplitude jitter. There was some hesitation with random loudness variations. Two levels of floor variations of 60 and 90 were studied. At the higher end, crackling distortion was present. At the 60 level, the irregularity did not create perceptions different from the ones discussed above.

Following these observations, we chose to concentrate on the half-life variations for the current attensons.

The Generation of a New Set of Threat Warnings

Research by the DERA and APU has shown that increasing the number of auditory warnings beyond six or seven becomes counterproductive (Patterson, 1990). Moreover warnings which have simple temporal patterns are confused (ibid.). Anticipating the need for threat warnings, the DERA has produced a four-level structure for warning sounds to ensure correct coding of urgency. Level I has the lowest urgency and level IV the maximum urgency. The levels for radar threats are as follows:

Level I	Undetected Search (Radar)
Level II	Acquisition by Radar
Level III	Tracked by Radar
Level IV	Missile or Gun Launched

The structure was implemented, as specified by DERA, as follows:

Level I	advisory attenson	+ voice message
Level II	dedicated attenson	+ 'Missile'
Level III	dedicated attenson	+ 'Unknown'
Level IV	dedicated attenson	+ 'Laser!'

For level I, level II and level III threats, we implemented a radar-like sweeping sound. Back-to-back damped and ramped envelopes produce a sound with asymmetry that can be controlled in a useful way, but the sharp peak where the two components of the envelope meet produces a sharp click in the sound. This suggested that the most appropriate envelope shape would be the rounded exponential (roex). The idea of the roex envelope came from work in the spectral domain on auditory filter shapes (Patterson, 1976). The rounded exponential (roex) shape has a rising exponential onset, a rounded top and a decaying exponential offset. The rounded top was introduced to make the filter flat at its centre frequency. When translated to the time domain the rounded top prevents the unpleasant click just as it makes the derivative of the filter function smooth at its centre frequency. The ramped half-life and the damped half-life of the roex are independent parameters. Hence we could produce smooth-sounding, time asymmetric envelopes – the central theme of these new sounds. To make a low-urgency sound, a roex envelope with long half-lives was applied to a low-pitched carrier.

Increasing the carrier frequency, decreasing the roex half-lives, and repeating the pulses rapidly generated warnings with greater urgency. So with the same type of attenson, changing one or two parameters changes the urgency (Patterson, 1990).

A prototype set of threat warnings was generated and recorded. The carrier used with the roex envelopes was IRN. Level I, *undetected search*, was made with a roex envelope with period 800ms, ramped half-life of 64ms and damped half-life of 128ms, with six seconds of silence between the pulses (in total, four pulses). The carrier was an IRN with 16 iterations and a 16ms delay (low pitch).

Level II, *acquisition by Radar*, was made with a roex envelope with period 400ms, ramped half-life of 32ms and damped half-life of 64ms, with three seconds of silence between the pulses (in total, four pulses). The carrier was an IRN with 16 iterations and an 8ms delay.

Level III, *tracked by Radar*, was made with a roex envelope with period 200ms, ramped half-life of 16ms and damped half-life of 32ms, with no silence between the pulses (in total, four pulses). The carrier was an IRN with 16 iterations and a 4ms delay.

Two level IV sounds were made, one for the *missile* message and the other for the *gun* message. Level IV, *urgent missile*, was made as a two-part warning. The first part was a dedicated attenson constructed with a ramped envelope with period 90ms, 30 per cent jitter in pulse time and a carrier of IRN with 16 iterations and a 2ms delay (high pitch). The second part, which identified the missile, was made of a long-period ramped sound (180ms) and an IRN carrier having an 8ms delay.

Level IV, *urgent gun*, was also made as a two-part warning. The first part was a dedicated attenson constructed with a ramped envelope with period 90ms, 30 per cent jitter in pulse time and a carrier of IRN with 16 iterations and a 2ms delay (high pitch). The second part, identifying a gun, was made with a long-period damped sound (180ms) and a random-phased harmonic carrier.

The radar sounds are clearly related in their form and sound quality and, as a result, they function as one warning with varying levels of urgency. This design can be expected to reduce the memory load for auditory warnings and increase the recognisability of radar warnings, at one and the same time.

Conclusions

The advanced audio displays being developed for aircraft require localisable auditory warnings with significant high-frequency energy in the region 6–12kHz. We have developed three methods of increasing the frequency range of existing warning sounds without changing the basic structure or sound quality by which they are currently identified. Basically we produced a high-frequency version of the warning sound and then blended the original and the high-frequency version together while ensuring that the high-frequency portion was sufficiently intense to support out-of-head localisation, and not so intense as to alter the essential character of the sound. In one case we extracted the envelope of the existing warning and filled it with a set of high-frequency sinusoids. In the second we extracted the spectral reflection of the sound above the Nyquist sampling frequency. In the third case, we replaced each cycle of the wave with two cycles of half the duration but the same basic shape. The three techniques were applied to each of the existing warning sounds and the best chosen by a committee of listeners. The high-frequency energy adds a sharpness or whistling to the sounds which makes them more arresting and identifiable.

We also developed a set of four related radar warning sounds to signal 'undetected, radar search sweep', 'acquisition by radar', 'being tracked by radar' and 'missile launched/gun fired'. The warning sounds were synthesised in the time domain using the results of new research on the perception of temporal asymmetry. The traditional radar sweep is emulated with a sound whose envelope is a pair of back-to-back exponentials with a rounded peak. A distinctive sound is produced by balancing the perceived duration of the rising and falling portions of the perception. This occurs when the half-life of the falling exponential is four times longer than the exponential of the rising exponential!

By varying the temporal parameters of the envelope in a systematic way, we can increase the perceived urgency of the warning. A synthetic carrier with a controllable degree of regularity gives the sweep a distinctive semitonal quality that can be intensified for the more urgent versions of the warning. These radar sweep sounds are clearly related in their form and sound quality and, as a result, they function as one warning with varying perceived urgency.

Acknowledgments

This research was supported by DERA contract No. ASF/3208 to the MRC Applied Psychology Unit. The chapter was originally presented in a somewhat different form at the 82nd AGARD meeting on 'Audio effectiveness in aviation', Copenhagen, October 1996.

References

Irino, T. and Patterson, R.D. (1996) 'Temporal asymmetry in the auditory system', *Journal of the Acoustical Society of America*, **99**, 2316–31.

Lower, M.C., Patterson, R.D., Rood, G., Edworthy, J., Shailer, M.J., Milroy, R., Chillery, J. and Wheeler, P.D. (1986) 'The design and production of auditory warnings for helicopters 1: The Sea King', Report AC527A, Southampton: Institute of Sound and Vibration Research.

Patterson, R.D. (1976) 'Auditory filter shapes derived with noise stimuli', *Journal of the Acoustical Society of America*, **59**, 640–54.

Patterson, R.D. (1982) 'Guidelines for Auditory Warning Systems on Civil Aircraft', CAA paper 82017, London: Civil Aviation Authority.

Patterson, R.D. (1990) 'Auditory warning sounds in the work environment', *Philosophical Transactions, Royal Society of London*, **B 327**, 485–92.

Patterson, R.D. (1994a) 'The sound of a sinusoid: Spectral models', *Journal of the Acoustical Society of America*, **96**, 1409–18.

Patterson, R.D. (1994b) 'The sound of a sinusoid: Time-interval models', *Journal of the Acoustical Society of America*, **96**, 1419–28.

Patterson, R.D. and Datta, A.J. (1994) 'Extending the frequency range of existing warning sounds, AAM HAP Auditory Warnings: Progress Report No. 1', APU Contract Report, Cambridge.

Patterson, R.D., Allerhand, M. and Giguere, C. (1995) 'Time-domain modelling of peripheral auditory processing: A modular architecture and a software platform', *Journal of the Acoustical Society of America*, **98**, 1890–94.

Patterson, R.D., Edworthy, J., Shailer, M.J., Lower, M.C. and Wheeler, P.D. (1986) 'Alarm sounds for medical equipment in intensive care areas and operating theatres', Institute of Sound and Vibration Research Report No. AC598.

PART III
AUDITORY COGNITION

5 The Interpretation of Natural Sound in the Cockpit

JAMES A. BALLAS, *Naval Research Laboratory, Washington DC*

Introduction

To paraphrase a popular song, the cockpit is alive with the sounds of events. Most interest has focused on the speech in the cockpit, since this indicates the interpretations and the actions of the crew. But, as Hutchins (1995) points out, the cockpit is a sociotechnical system that represents information both internally and externally and, whereas the speech events might represent the internal events of the crew's cognitive processing, other sound can represent the events of the system. Auditory alerts are an explicit representation of a subset of these events. The purpose of this chapter is to discuss how sounds other than the engineered auditory alerts and crew speech represent information in the cockpit. For lack of a better term, this is referred to as 'natural' sound, as, for example, the sounds naturally produced by system events, such as mechanical failures. The chapter presents an analysis of the occurrence of these types of sounds in accident reports. It turns out that these sounds are often brief, and so it is useful to understand the factors that affect the rapid and accurate interpretation of short sounds (Ballas, 1993a). Here additional findings on the perception of temporal and spectral features seen in spectrograms are discussed. Finally a general framework developed by Guyot (1996) is presented which is useful in characterising the cognitive processes in sound interpretation. This type of framework will be important as one adopts a general model of sound perception in an applied setting such as an aircraft cockpit.

The crash of Delta Flight 1141, on 31 August 1988, provides two examples of sound interpretation in the cockpit. The first example is one of sound misinterpretation. The US National Transportation

Safety Board (NTSB) found that the flaps had been set incorrectly on takeoff, resulting in an unusually high pitch angle for the aircraft. This disrupted the air flow into the engines, causing a compressor failure, as described in the following excerpt:

> Two seconds after takeoff, a snap was heard, and then the control column began to shake. 'Something's wrong,' Davis [Capt. Larry Davis, who was in control of the aircraft] said. Two seconds later, the first of five engine compressor stalls was heard, indicating a disruption of airflow into the engines. At the time, the pilots believed the sounds signaled an engine failure. Copilot Carey W. Kirkland called that out after the second sound. Then in rapid succession three other compressor stalls, which sound similar to a car backfiring, were heard on the tape. 'We've got an engine failure,' said one of the three pilots, whose voice was unidentifiable on the tape. (*Washington Post*, 1 November 1988, p.A3)

The pilot, believing an engine had failed, hesitated in initiating full power. This decision was based on a misdiagnosis of the engine state and a misinterpretation of the compressor stall sound. A second example of sound interpretation in this accident comes in the post-accident analysis. At issue was whether the flaps had been set correctly. The analysis included a simulation to determine the source of two clicks that were picked up by the cockpit voice recorder, just after the pilots had been given instructions to taxi to the runway. Attempts to match these clicks to those made by setting the flap handle were not successful.

References to Heard Sounds in Accident Briefs

Incidents such as this prompt one to think about what role sound interpretation has played in accidents. In order to address this issue, an analysis of accident report briefs was conducted. The source of this analysis was an on-line database of the NTSB Accident Investigation Dockets. The on-line version of these dockets contains a condensed summary of the basic facts, including a narrative and a probable cause. These are made available after official approval of the NTSB. The briefs have been available electronically since 1994. Several keyword searches of the database were performed. The results reported here are based on an analysis of 80 briefs which included the word 'noise' in the narrative of the accident. These incidents will be referred to as 'Heard-a-Sound' incidents.

The sounds referred to in the briefs were typically produced by the engine or a mechanical failure in the aircraft structure. Since some of the incidents involved helicopters, some of the noises involved failures in the drive train. The noises were described using words such as 'bang', 'splutter' and 'pop', with occasional further elaboration. A few incidents described the offset of a sound, especially the cessation of engine noise. Occasionally an explicit mention of 'normal' operating noises was made. The sounds were heard by either crew or ground witnesses. Most of the incidents involved general aviation. None of the briefs mentioned a cockpit voice recorder.

Accident Cause for Heard-a-Sound Incidents

The cause was encoded for these 80 incidents using the following categories: pilot, aircraft, maintenance, other and unknown. Only the primary cause listed in the NTSB brief was encoded. The causal attributions in these briefs represent the official NTSB position. Two distributions of causes were examined: crew Heard-a-Sound (Figure 5.1) and witnesses Heard-a-Sound (Figure 5.2). The salient difference in these two distributions was the significant ($z = 2.63$, $p < 0.05$) difference in the human error proportion. The distribution in

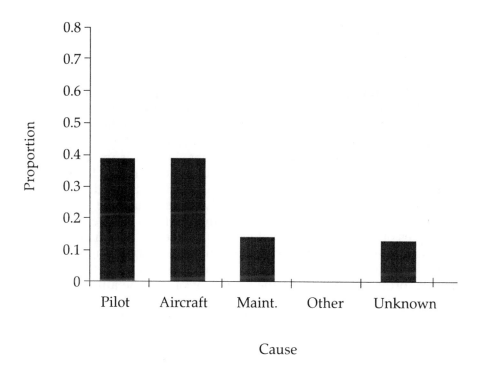

Figure 5.1   Causes of accidents with a noise heard by the pilot

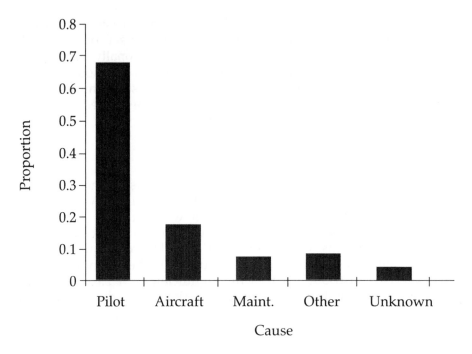

Figure 5.2   Causes of accidents with a noise heard by a witness

Figure 5.2, for witnesses Heard-a-Sound incidents, is the more typical result. The human error proportion in this figure is around the two/thirds to three/quarters that is typically reported (for example, Nagel, 1988). Evidently, in incidents in which the crew heard a sound, the accident was more likely to be a mechanical failure than is typically found. This has two implications. First, it is important that sound interpretation be accurate, because the sound is likely to be indicative of the cause of the impending accident. Second, the sound and the incident producing it are likely to be a surprise since these incidents are not precipitated by human actions. There are many examples of incidents precipitated by human error in which the crew had some foreboding that the situation was deteriorating. A classic example is the analysis of the voice recording in the Tenerife accident. This recording showed that the copilot of one of the jumbo jets was anxious about initiating the takeoff. The aircraft subsequently collided with a second jumbo jet which was taxiing on the same runway.

  This type of analysis could help design a sound identification module for pilot training programmes. It would be important to further specify the types of sounds, and the types of incidents which could cause them. In particular, if alternative events could cause similar sounds, these should be further specified. The common use of general descriptors such as 'bang' suggests that the pilots in these incidents sometimes extracted very little information from the sound. Training materials would need to be developed for such a module. One of the few examples of such materials was prepared for motor mechanics (Weissler, 1986). The developer of these materials visited car repair facilities to obtain recordings of malfunctioning engines. The recordings were then filtered to isolate the key sound event for diagnostic purposes. The final training tapes included recordings which contrasted the isolated sound with the original recording so that the mechanic could learn to focus on particular aural cues upon hearing the total suite of sound.

Linguistic Functions of Heard Sounds

The sounds heard in these incidents conveyed different types of information, and one of the challenges is to be able to classify and describe the different forms. As Stanton and Edworthy point out in Chapter 6 of the present volume every sound has a potential function. One challenge is how to describe and classify the function. Much of the work in auditory alerts is to design and evaluate sounds which signify specific critical events. However 'signify' is a very general term and it would be useful to be able to expand the notion of functionality. For this type of analysis it is useful to adopt the position that the aural environment is a communication system (Truax, 1984). Within this communication system, non-speech sound can achieve specific linguistic functions (Ballas, 1993b). Five linguistic functions were used to analyse the references to noises in the accident briefs.

1   *Exclamation*: a sudden vehement utterance. For this analysis, the focus was on the alerting aspect of exclamation, asking whether the sound captured attention. Although every sound is alerting by its nature, sounds were coded only if a phrase or word in the narrative (such as 'loud') explicitly mentioned that the sound probably could not have been ignored.
2   *Deixis*: a demonstrative such as the words 'there', 'here', 'this', 'those', and 'now'. This function was present whenever the sound directed attention towards a specific event, object or process (for example, 'noise from the engine').
3   *Simile*: a figure of speech in which two things are explicitly compared using a word such as 'like' or 'as'. In this analysis, the sound functioned as a simile whenever the listener used it to interpret the status or state of an event or process (for example, 'ran up the engine until the popping sound stopped').

4   *Metaphor*: a figure of speech in which one notion is said to be identical to another, dissimilar notion. To qualify as a metaphor, the sound had to be described as another sound-producing event that was different from what was actually causing the sound (such as hearing a noise similar to a 'rifle being fired').
5   *Onomatopoeia*: an utterance whereby the sound itself symbolises the meaning. In this case, the sound was described with a phrase that represented the acoustics of the sound itself (such as 'bang').

The distinction between simile and metaphor is subtle, and it is legitimate to ask whether this distinction is worth maintaining. The general issue of how non-speech sound can represent information is being actively pursued, and other definitions of simile and metaphor have been proposed (see Kramer, 1994). The distinction in this analysis was used because one function (simile) denotes that a listener is using a sound to interpret the state of a process that is probably unseen. The other function (metaphor) denotes that a listener is describing the sound itself by referring to another type of sound-producing event. The use of this latter function means that the listener wants to describe the acoustics of the sound in familiar terms.

Each instance was coded with the linguistic functions the sound achieved. Sounds were classified in as many functions as they achieved. In addition, if the description included the specific aural properties of the sound, this was also noted. As before, the results were analysed according to those who heard the sound, crew (Figure 5.3) or witness (Figure 5.4).

The predominant function of these sounds for both crew and witnesses was exclamatory (that is, alerting) and the proportion for this function was significantly greater than the next highest for both pilots ($z = 4.26$, $p < 0.05$) and witnesses ($z = 2.95$, $p < 0.05$). Crews described the sounds this way more than witnesses, but not significantly so. Crews described sounds as more onomatopoeic and deictic, and these differences were nearly significant ($z = 1.82$, $p = 0.068$, and $z = 1.67$, $p = 0.095$). It would be useful to investigate further these types of differences to determine whether they result from the task demands or are due to the accident investigation process. The pilot's task is to quickly identify the source of any malfunction, so the acquisition of any deictic information would be important. The somewhat greater proportion of deictic interpretation by the crew implies that the crew use the sound as a reference to specific system components. It would be interesting to find out if the accident investigation itself considers the pilot's interpretation of sound as part of the causal analysis. Since the witnesses are being interviewed to gather information about the sequence of events, they might be encouraged to describe the inferences they made upon hearing the sound.

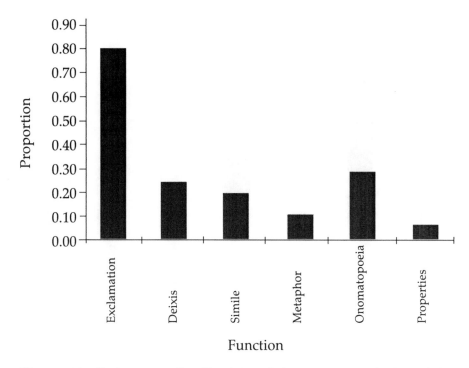

Figure 5.3   Frequency distribution of the way sounds heard by pilots functioned

Psychoacoustics of Identifying Brief Sounds

Some examples of the sounds mentioned in the narratives are the following:

- 'loud thump' (NTSB File #1925)
- 'two bang noises' (NTSB File #1423)
- 'loud crashing noise' (NTSB File #1295)
- 'puffing or popping noises' (NTSB File #469)

One can assume that these particular sounds were brief. Other descriptions referred to longer sounds, such as continuing engine noise, and a few descriptions explicitly mentioned the time course of the sound (for example, a 'continual rumbling noise', NTSB File #791). The brief sounds mentioned in the first kind of descriptions are the type of sounds that were the focus of several experiments reported elsewhere (Ballas, 1993a; Ballas and Mullins, 1991). The results indicated that multiple factors affect the identification of brief, everyday sounds. The factors include causal uncertainty (the degree

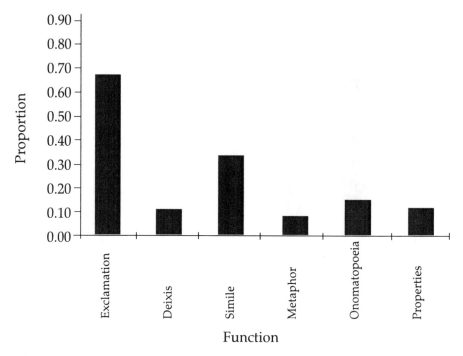

Figure 5.4   Frequency distribution of the way sounds heard by witnesses functioned

to which multiple mappings exist between an acoustic pattern and different types of causal events), ecological frequency (how often the sound actually occurs), stereotypy (whether the sound matches a mental stereotype) and certain acoustic properties of the sounds. The latter factor is the focus of further examination in this chapter.

Scope of Psychoacoustic Analysis

A set of 41 sounds (Figure 5.5) was investigated in the previous research. Acoustic analyses examined the overall properties such as duration, average magnitude, peak magnitude, power, Fast Fourier Transform (FFT) spectrum, and 1/3 octave spectrum, and moments of the FFT spectrum, computed following the work by Chen (1983). A principal components analysis of the 1/3 octave band spectra extracted four factors corresponding to four frequency bands. Further details of these analyses and the results are available in Ballas (1993a). Here, the focus will be on further analysis of the spectrograms, which are illustrated in Figure 5.5. The spectrograms are in the top part of

Sound and description	H_{st}	Count	Spectrogram
1 Telephone ring high-pitched ringing	yes	11	
2 Clock ticking three ticking sounds	yes	3	

3 Car horn
   single medium-
   pitched horn
	yes	na	

4 Doorbell
   two chimes; first
   higher pitched
	yes	2	

5 Automatic rifle 5 shots	yes	5	

6 Boat whistle
   medium-pitched
   whistle
	yes	na	

Sound and description	H_{st}	Count	Spectrogram
7 Water drip single water drip	no	–1	
8 Bell buoy one medium- pitched bell	yes	na	
9 Foghorn low-pitched horn	yes	na	
10 Water bubbling continuous, bubbling sound	no	–7	
11 Bugle 5 notes; pitch increasing for notes 1–4	yes	5	
12 Gunshot indoors single shot, no echo	no	–1	

Sound and description	H_{st}	Count	Spectrogram

13 Lawn mower
    continuous
    combustion noise
	yes	na	

14 Church bell
    two high-pitched
    bells
	yes	2	

15 Oar rowing
    sound of water
    flowing
	no	−1	

16 Door knock three rapid knocks	yes	3	

17 Toilet flush
    toilet flushing
    water
	no	−3	

18 Footsteps two rapid footsteps	yes	2	

Sound and description	H_{st}	Count	Spectrogram
19 Fireworks explosion	no	−2	
20 Cigarette lighter lighter lit, quick noise burst	yes	1	
21 Touch tone single tone	yes	na	
22 Door opened metallic door-latch opening	no	−1	
23 Bacon frying sound of frying oil	no	na	
24 Hammering two quick taps	yes	2	

Sound and description	H_{st}	Count	Spectrogram
25 Sub dive horn pitch increasing and decreasing	yes	na	
26 Clog footsteps three impact sounds	yes	3	
27 Car ignition three engine rotations	yes	3	
28 Tree chop single chop	no	−1	
29 Power saw whirring sound increasing in pitch	no	na	
30 Door latched two partly muffled latching operations	no	−5	

Sound and description	H_{st}	Count	Spectrogram
31 Cork popping single pop	no	−1	

32 File drawer closed
      metallic wheeling-
      drawer closing
	no	−3	

33 Door closed
      two muffled impact
      sounds
	no	−3	

34 Car backfire
      one explosive
      backfire
	no	−1	

35 Jail door closed
      heavy metallic door
      shut with echo
	no	−2	

36 Gunshot outdoors
      single shot with
      echo
	no	−3	

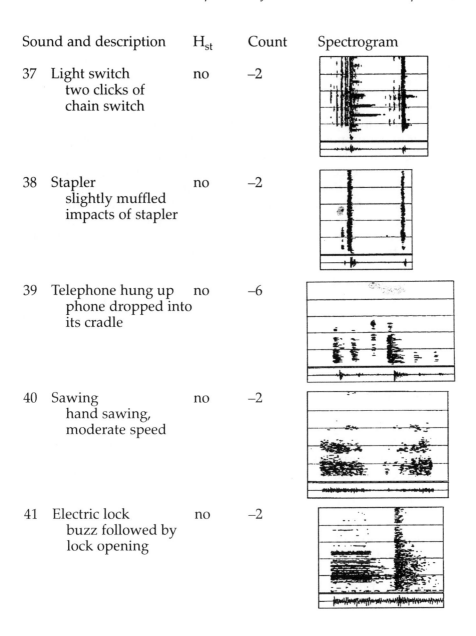

Sound and description	H_{st}	Count	Spectrogram

37 Light switch
    two clicks of
    chain switch
	no	–2	

38 Stapler
    slightly muffled
    impacts of stapler
	no	–2	

39 Telephone hung up
    phone dropped into
    its cradle
	no	–6	

40 Sawing
    hand sawing,
    moderate speed
	no	–2	

41 Electric lock
    buzz followed by
    lock opening
	no	–2	

Figure 5.5 Spectrograms, spectral/temporal entropies (H_{st}) and component counts for 41 sounds

each panel in the figure; the waveform is in the bottom part. The spectrograms show that the sounds contain a variety of features including harmonics, continuous spectral bands, bursts that are similar spectrally, and spectral shifts similar to the formants in speech.

Several temporal properties were calculated, including the number of components in the sound, the duration of the components and the ratio of component duration to total duration. These calculations defined the components in terms of amplitude decrease for a minimum period of time. However these calculations produced results that were sometimes inconsistent with a visual examination of the spectrograms, so the latter was also used. The spectrograms were visually examined for other properties besides components. Combinations of the spectral/temporal properties were also included in the analyses. In particular, the Boolean union of two properties emerged as a factor related to perceptual performance. This union was harmonics in continuous sounds, or similar spectral patterns in the bursts of non-continuous sounds. This union describes a form of spectral/temporal entropy in that it refers to the repetition of spectral information over time. It will be referred to as H_{st}.

Spectral/Temporal Entropy: H_{st}

The correlations of H_{st} and the time to identify the sounds ($r = -0.59$) and accuracy ($r = 0.57$) were significant ($p < 0.05$). Sounds which had one of the two forms of H_{st} are indicated in Figure 5.5. The judgments listed here were made by a linguist who was not familiar with the actual sounds or the nature of the research. The spectrograms themselves are listed according to identification time. Clearly the presence of H_{st} diminishes as you scan through the spectrograms. Specifically most of the first 10 sounds have one of the two forms of H_{st}; none of the last 14 sounds has either one.

This result is interesting because of the parallels between H_{st} and the acoustic properties that are emerging in research into urgency encoding. Edworthy *et al.* (1991) have identified a set of spectro/temporal properties that are related to the perception of urgency, and Haas and Edworthy (1995) have examined the relationship between some of these properties and performance. The sounds are constructed according to Patterson's principles and include a pulse as the elementary unit of sound. The pulse is constructed from a harmonic series and has a consistent envelope. These design principles prescribe sounds that will have a similar spectral pattern in the bursts, a property which is similar to H_{st}. Thus there is evidence that accurate identification of everyday sounds is related to the same types of acoustic properties that are being studied and validated for the design of synthetic alerts.

Component Count

One of the properties that has seen little investigation in the studies of urgency encoding is the number of components or pulses. Visual examination of the spectrograms suggested that this property might be related to identification performance. In order to investigate it, the number of pulses was determined for each sound by visual examination of the spectrograms, and was weighted by whether the sound had H_{st}. This weighting was a dummy variable, coded as +1 if the sound had H_{st}, negative otherwise. The intention was to determine whether the number of components was both positively and negatively related to identification performance, depending on whether the components were spectrally similar. Continuous sounds were excluded from this analysis, so only the presence of similar spectral patterns in the repeated bursts was relevant to determining H_{st}. The particular coding is shown in Figure 5.5 with the spectra so the reader can judge the validity of the encoding. Similar results were found when the number of components was computed as described previously. The weighted variable was significantly correlated ($r = 0.60$, $p < 0.05$) with the time to identify the sound. As shown in Figure 5.6, there is a relationship between component count and identification time, but the form of this relationship depends on whether the components are a repeated spectral pattern (positive component count) or a series of variable spectral components (negative component count). Simply stated, this means that, with everyday sounds, the repetition of a component improves identification, whereas the aggregation of different components impairs identification. This result holds for sounds that are relatively brief (<700 ms). But a series of studies which examined the identification of a sound embedded in a longer sequence of other sounds supported the conclusion that identification is compromised when the series of sounds is highly varied (Ballas and Mullins, 1991).

A General Model of Everyday Sound Perception

As noted previously, several factors have been implicated in the identification of everyday sound perception, including causal uncertainty, ecological frequency, stereotypy and certain acoustic properties. Several of these factors imply that cognitive processes are operating on a memory of sounds that is structured to represent

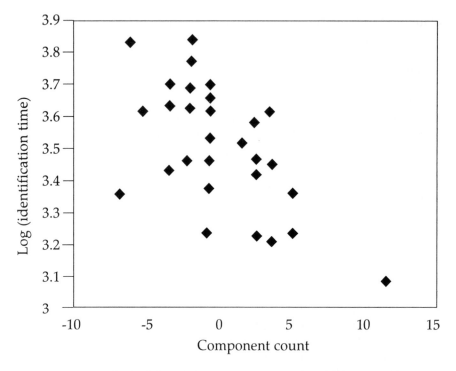

Figure 5.6   Relationship between component count, weighted by H_{st}, and identification time in log units

classes or types of sounds. The effects of causal uncertainty are consistent with a perceptual abstraction process which handles the variation across the instances of a particular type of sound-producing event. For example, variation in the way a stapler is pressed (Ballas, 1994) or how footsteps are generated (Li *et al.*, 1991) does not preclude the accurate perception of these sounds, even though this variation will generate very different waveforms. However the abstraction process can also result in interpreting a stapler press as a pull-chain light switch. Guyot (1996) has recently investigated the cognitive structure of everyday sound perception in experiments which examined the categorisation process. Her research suggested that the abstraction process within categorisation operates at three levels. At one level the type of excitation (mechanical or electronic) is identified. At the next level, the movement producing the acoustic pattern is identified. At the third level, the event is identified. She focused at the second level, and described the perceptual properties for three forms of gestures: scratching, rubbing and grinding. She also developed an overall model, which illustrates the processes involved (Figure 5.7).

This model has two processes that produce observable behaviour, recognition and qualitative appreciation. Both of these processes follow from a categorisation process. The important implication is that sound recognition and qualitative assessment is driven by a cognitive process which refers to some form of mental representation and abstraction for sounds. Guyot's research has suggested that categories represent the type of sound-producing gesture. Other categorisation schemes have been presented by Schafer (1977), Gaver (1993) and Ballas (1993a).

Schafer's taxonomy is based upon references to sound in published writing and emphasises the type of human involvement in the sound-producing process. The categories he proposed were natural sounds, human sounds, societal sounds, mechanical sounds, quiet/silence and sounds as indicators, including horns, clocks, telephones and warnings. Gaver's taxonomy follows the direction initially proposed by Gibson (1966), who suggested three classes of sounds: solids, wind and water. Gaver characterised these as vibrating objects, aerodynamic sounds and liquid sounds. Each of these was further specified. Vibrating objects decomposed into impacts, scraping and others; aerodynamic sounds into explosions and continuous wind; liquid sounds were either dripping or splashing. He further showed how a complex aural event, such as a motorboat, can be described with the elements within his taxonomy. Finally Ballas (1993a) examined the clustering of the 41 sounds using factors computed from perceptual and cognitive ratings and resulted in four classes of sounds: (1) sounds that are produced with water (drip, splash, bubble, flush) or in a water context (boat whistle, foghorn); (2) signalling sounds (telephone, doorbell, bugle, submarine dive horn and carhorn), and sounds that connote danger (fireworks, auto, rifle and power saw); (3) door sounds and sounds of modulated noise; and (4) sounds that have two or three transient components. These classes overlap some of those proposed by Schafer and Gaver, and the last two have aspects that are consistent with the concept of gestures in Guyot's approach.

The commonalities among these four categorisation schemes provide an encouraging direction for further research. A general model such as Guyot's can be used as a common frame of reference for this research as well as studies into auditory alerts. For example, Hellier *et al.* (1995) investigated the semantic interpretation of simple sounds, focusing on the issue of parameter mapping. They referred to two types of meaning in the mapping of acoustic parameters to semantic interpretations. The first is the interpretation of parameter change, particularly whether the change in a parameter will reliably elicit a specific change in interpretation. The second is the achievement of semantic constancy despite a change in acoustics. The

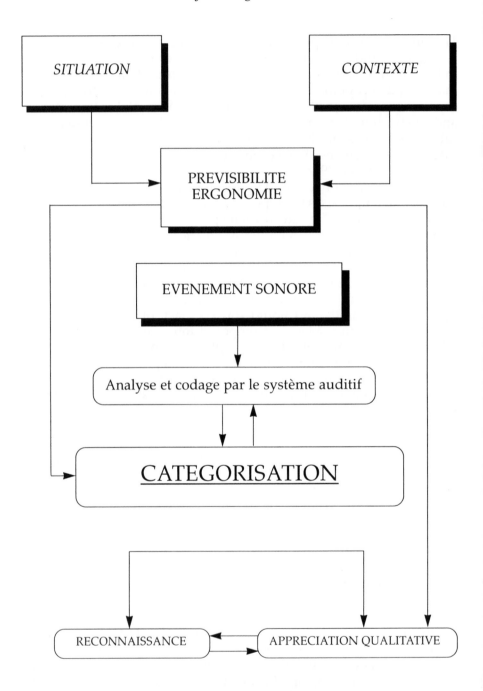

Figure 5.7   General model of the perception of environmental sounds, illustrating the processes involved

Source: Guyot (1996).

former would suggest a categorisation process that maps particular forms of dynamic acoustics to a semantic category. The latter suggests that some categories of sounds are abstractions across a range of acoustic properties. Categorisation can encompass both forms. Most of the judgments that Hellier *et al.* used were some form of qualitative appreciation.

In this model, context and the situation influence the categorisation process (and thereby both recognition and qualitative appreciation) but can also influence qualitative judgments directly. A variety of ecological studies have shown these types of effects. Southworth (1969) found that judgments of sound depended upon the characteristics of the site, the sound itself and the intensity of the sound. Anderson *et al.* (1983) found that appreciation of a site depended upon the meaning of the sounds at the site and the appropriateness of the sound for the site. Natural sites were rated more positively if the sounds were also natural, and less positively if the sounds were urban. Jenkins (1985) reported that blindfolded students would discover auditory landmarks within a building, and that noises, resonances and echoes would provide information about the size of a room. Truax (1984) describes analyses of the soundscape of cities and villages, using the term 'soundscape' to refer to the way the environment is understood by the listeners within it. This chapter extends our understanding of the cockpit 'soundscape' and how its listeners interpret its 'natural' sound.

Acknowledgments

The preparation of this chapter was supported by the Office of Naval Research. The author wishes to thank Steve Shamblen for providing a pilot's perspective on sound in the cockpit and for reviewing a draft of this chapter.

References

Anderson, L.M., Mulligan, B.E., Goodman, L.S. and Regen, H.Z. (1983) 'Effects of sounds on preferences of outdoor settings', *Environment and Behavior*, **15**, 539–66.
Ballas, J.A. (1993a) 'Common factors in the identification of brief, miscellaneous, everyday sounds', *Journal of Experimental Psychology: Human perception and performance*, **19**, 250–67.
Ballas, J.A. (1993b) 'Interpreting the language of informational sound', *Journal of the Washington Academy of Sciences*, **83**, 143–60.
Ballas, J.A. (1994) 'Effect of event variations and sound duration on identification of everyday sound', in G. Kramer and S. Smith (eds), *Proceedings of the Second International Conference on Auditory Display, ICAD '94*, 7–9 November, Santa Fe, New Mexico.
Ballas, J.A., and Mullins, R.T. (1991) 'Effects of context on the identification of everyday sounds', *Human Performance*, **4**, 199–219.

Chen, C.H. (1983) 'Pattern recognition processing in underwater acoustics', *Pattern Recognition*, **16**, 363–640.

Edworthy, J., Loxley, S. and Dennis, I. (1991) 'Improving auditory warning design: relationship between warning sound parameters and perceived urgency', *Human Factors*, **33**, 205–32.

Gaver, W.W. (1993) 'What in the world do we hear? An ecological approach to auditory event perception', *Ecological Psychology*, **5**, 1–29.

Gibson, J.J. (1966) *The Senses Considered as Perceptual Systems*, New York: Houghton Mifflin.

Guyot, F. (1996) 'Etude de la perception sonore en termes de reconnaissance et d'appréciation qualitative: une approche par la catégorisation', Paris: Universite de Paris VI; Laboratoire d'Acoustique Musicale.

Haas, E.C. and Edworthy, J. (1995) 'The perceived urgency and detection time of multitone auditory signals', *Proceedings of Human Factors in Alarm Design II*, Chilworth Manor, University of Southampton.

Hellier, E., Edworthy, J. and Hards, R. (1995) 'An investigation into the "meaning" of simple sounds', *Proceedings of Human Factors in Alarm Design II*, Chilworth Manor, University of Southampton.

Hutchins, E. (1995) 'How a cockpit remembers its speeds', *Cognitive Science*, **19**, 265–288.

Jenkins, J. (1985) 'Acoustic information for objects, places and events', in W. Warren and R. Shaw (eds), *Persistence and Change*, Hillsdale, NJ: Erlbaum Associates.

Kramer, G. (ed.) (1994) *Auditory Display: sonification, audification and auditory interfaces*, Santa Fe Institute Studies in the Sciences of Complexity, Proc. Vol. XVIII, Reading, MA: Addison-Wesley.

Li, X., Logan, R.J. and Pastore, R.E. (1991) 'Perception of acoustic source characteristics: Walking sounds', *Journal of the Acoustic Society of America*, **90**, 3036–49.

Nagel, D.C. (1988) 'Human error in aviation operations', in E.L. Wiener and D.C. Nagel (eds), *Human Factors in Aviation*, San Diego: Academic Press.

Schafer, R.M. (1977) *The Tuning of the World*, New York: Knopf.

Southworth, M. (1969) 'The sonic environment of cities', *Environment and Behavior*, **1**, 49–70.

Truax, B. (1984) *Acoustic Communication*. Norwood, NJ: Ablex.

Weissler, P. (1986) 'Strange engine sounds', *Home Mechanix*, **82** (May), 82.

6 Auditory Warning Affordances

NEVILLE A. STANTON, *University of Southampton* and **JUDY EDWORTHY,** *University of Plymouth*

Introduction

Many writers have announced their dissatisfaction with the design of alarms and warnings. Some even propose new ways of presenting the information. For example Patterson (1982) has advocated the use of small units of sound (pulses) from which bursts of sound lasting one or two seconds are constructed. The bursts are made into complete warnings by interspersing the bursts with silence. This method of sound construction has been used in some design applications, for example in aviation (Lower *et al.*, 1986) and the hospital environment (Patterson *et al.*, 1986). Stanford *et al.* (1988) have advocated the use of alarms constructed from short bursts of vowel-like sounds, which have been shown to be aesthetically preferable to the traditional warnings.

However, as yet, no one has seriously investigated the possibility of making alarm sounds that are representative of the event to which they are alarming, although Gaver (1989) has advocated the use of environmental sounds in the development of sound interfaces for computers. Even in such applications, researchers have not yet taken the step towards exploring the relationship between the end-user's existing knowledge of sounds and their meanings in the development of warning sets.

In a user-centred approach to warning design, the authors have proposed the adaptation of a standardised methodology for developing and testing information symbols (ISO/DIS 7001) for auditory warning design. The underlying principles of the procedure have been explained in previous publications (Stanton and Edworthy, 1994; Edworthy and Stanton, 1995). In the current chapter, the underlying theory of auditory affordances are explored in an Intensive Treatment Unit (ITU). This domain has been chosen for several reasons. First, the auditory warnings are often highly ambiguous, and problems have been identified in previous research (for example, Stanton, 1993). Second, the warnings may be communicating vital information, which could be of life or death significance. Therefore it is essential to maximise the recognition of these sounds. Finally the absolute number of warnings communicated is relatively small compared to other settings. Therefore it provides a non-trivial example without unecessary complexity.

Central to the methodology is the theory of auditory affordances. The theory of auditory affordances is based upon Gibson's (1979) theory of affordances. According to Gibson, we do not experience the world directly, but via an internal mental representation of the world. This view proposes that people are active participants in the world and direct their experiences of it. As Gibson proposed, the structures that enable us to interpret the world are developed through our own activities and experiences. Further Gibson claimed that what we attend to is not the purely physical aspects of objects in the world (such as size, shape or colour), rather we seek to interpret their meaning (they can be eaten, sat on, thrown, and so on). This approach to understanding human perception can be extended to perception of sound events, for example auditory warnings. We therefore propose a theory of auditory affordances. Put simply, we argue that an auditory warning is perceived in terms of its potential for action. This theory has four main propositions.

1   *We are surrounded by sounds*   Sounds form part of our everyday life. We may be unaware of just how important sound cues are to perform our activities: for example, the engine sounds that indicate it is time to change gear in a manual car, the clunk that informs us that the door has locked, and so on.

2   *We are introduced to sounds*   Sounds are not merely encountered, we are introduced to them. The function of a sound is explained to us when people teach us how to use an object.

3   *We learn about sounds through seeing other people respond to them*   We acquire our knowledge about how to respond to sound events by watching others. This can occur formally (that is, through training) or informally. The social aspect of affordances was emphasised by Gibson in his writings.

4   *A sound has a definite function*   Each sound has the potential to be ascribed to a function. At present the mapping of sounds to functions appears, at best, suboptimal and, at worst, random.

These propositions enable us to develop testable hypotheses for the basis of an experimental study. The basis for the theory of affordances is the idea of a direct linkage between the object (or sound) and action. The object (or sound) is perceived in terms of its possibility for a particular type of action. We have simply extended this notion to suggest that an auditory warning may be perceived in terms of its potential for action. For example, the sound from a syringe pump should confer the action of replacing the drug. The methodology that we have developed to assist in the design of warnings is described in the following section.

The Methodology

To design warnings based upon the theory of auditory affordances, we have developed the following methodology, as illustrated in Figure 6.1. The methodology involves asking people to propose potential warning sounds. This in turn allows us to investigate end-users' understanding of auditory affordances directly. The following sections illustrate the main aspects of the methodology.

Establish the Need for Warnings

First of all the warning referents (the situations to be represented by the warnings) need to be identified. It is vital to rationalise the operational situations for which the warnings will be needed and the set of referents need to be considered together. The output from this first phase is a set of functions for which warning sounds will be required.

Existing and Modified Sounds

Second, existing sounds are sampled from the environment. There may already be warnings for many situations which are deemed to be required. Once sampled, it is relatively easy to transfer the sounds to computer for editing and permanent storage.

Generate Trial Sounds

Apart from using existing and modified versions of warning sounds that are currently in use, both the designer and potential end-user are at liberty to generate ideas for new sounds. Trial sounds may be generated in a variety of ways, for example:

- by collecting ideas from the user population through brainstorming and interviews with a description of the referents to hand,

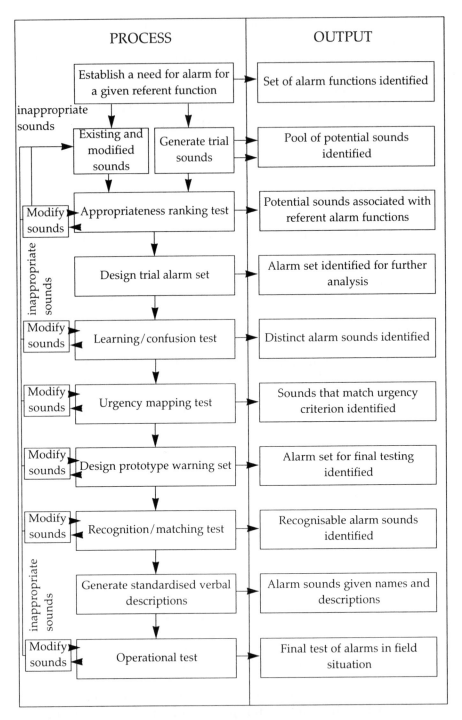

Figure 6.1   Procedure for designing auditory warnings

- sampling sounds from the natural environment to which the referents may relate,
- serendipitous discovery of sounds by allowing user groups to play with a synthesiser with specific referents in mind,
- sampling sounds from CD-ROMs.

Implementation of this methodology is possible using a particular equipment set-up. In order to develop and test new auditory warning designs, the following equipment is required:

- computer with sound card,
- synthesiser,
- midi interface,
- CD-ROM drive,
- CDs with sounds,
- sequencing software,
- sound-editing software,
- cassette recorder and player.

Figure 6.2 illustrates how these components may be linked together. The computer system in this configuration enables great flexibility in the production of auditory warnings, as required by the methodology (see: Generate trial sounds). The configuration of the equipment in Figure 6.2 enables existing sounds to be sampled from the environment (with the cassette recorder) and the sounds to be transferred to the computer. These sounds may be altered with the sound-editing software. In addition, new sounds may be selected from CD-ROM catalogues and likewise manipulated. New sounds may be generated on a synthesiser and edited on sequencing software. Thus one is able to select, sample and generate sounds to create a pool of potential auditory warnings. At this stage in the process, the main task is to generate as many sounds as is appropriate. No editing of ideas should be entered into, as this may dampen the creative process. All the sounds from the first three stages are entered into the first test (see the next section).

Appropriateness Ranking Test

In this stage each of the referents is presented in turn and participants (preferably an absolute minimum of 12 participants) are asked to rate the appropriateness of each sound to each of the referents. An example of the procedure follows.

1 Referent 1 is presented to a representative group of users. For example, the situation (such as a cardiac arrest) is presented.

Computer with sound card

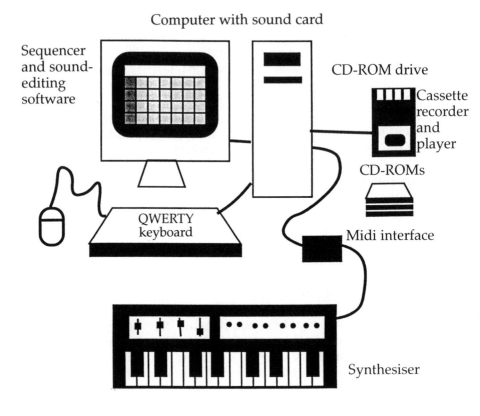

Figure 6.2   **Configuration of hardware and software**

2   The first sound is played and the users have to mark sound 1 on a seven-point Likert scale for appropriateness, where one is highly inappropriate and seven is highly appropriate. If they are unsure, the scale mid-point should be used. If users did not hear the sound properly it can be replayed. Each sound should be judged on absolute appropriateness, rather than relative to other sounds. Thus the user is making a judgment based upon his or her skills, experience and knowledge.

3   If there are more sounds for Referent 1, each is presented in turn and noted for appropriateness.

4   When the sounds for Referent 1 are completed, the next referent and its sounds are presented. This continues until all referents and sounds have been rated.

If any referent fails to achieve median ratings above the mid-point, the sounds may require modification or new sounds will need to be generated, and the testing process entered into again. When all referents have at least one sound above the mid-point for appropriateness the trial warning/alarm set can be designed.

Design Trial Warning/Alarm Set

Each referent is assigned a main warning sound and, if available, up to two reserve sounds. The main sound is the sound found to achieve the highest rating across all of the users' ratings. For example, a nomic representation for a specific referent (such as the sound of an actual heartbeat for the cardiovascular monitor) may be rated as the most appropriate sound and a metaphoric representation (such as simulated heartbeat sound) may be rated as the second most appropriate. (See Edworthy and Stanton, 1995, for a fuller description of these definitions.)

Learning/Confusion Test

There are two parts to this test. First, the learning test seeks to establish the 'learnability' of the sound. A paired-associate learning task may be used to determine the number of trials taken by participants to learn the sound for each referent. Second, during the learning process, instances of the confusion between sounds are recorded for investigation later. Confusions may occur through acoustic or semantic similarities. For example, warnings may be confused because they sound alike or they have similar functions irrespective of their sound quality.

Urgency Mapping Test

Before this test starts, expert judges will be required to rate the situational urgency of each of the referents. This rating will be compared to the urgency rating given by the participants for each of the sounds. As each of the sounds is played, the participant is required to make a judgment as to the urgency of that sound on a five-point Likert scale (from 'not urgent' to 'very urgent'). It is worth noting that urgency may be associated with the acoustic properties of the sound and/or the semantic content of the sound. It is important that all of these tests are carried out with the personnel who will use the resultant warnings.

Discrepancies between the situational urgency of the referent and the perceived urgency of the sound will require further exploration to see if these incompatibilities can be resolved. This may require some alteration of the acoustic qualities of the sound.

Design Prototype Warning Set

The prototype warning set is constructed on the basis of the modified sounds which have passed through the previous reiterative processes. This set enters into the final testing phase.

Recognition/Matching Test

The recognition/matching test involves the presentation of the warnings to participants, who are required to select a referent as each sound is heard. Their selection is 'with replacement', so that each referent is available for each sound. This avoids the problem of participants 'backing themselves into a corner' as the range of referents to choose from becomes increasingly restricted. Final modifications to sounds are permitted in this stage.

Generate Standardised Verbal Descriptions

On the basis of the last test, each of the sounds is given a verbal description. This serves simply for terms of reference and documentation purposes. There are two versions of the verbal description. The first provides a general, non-technical, description of the warning and the second provides a more detailed account of the acoustic structure.

Operational Test

The final test of the effectiveness of the auditory warning set is an operational test. This could be undertaken by playing the sounds randomly in the environment and collecting responses from potential end-users on which referents they believe the sounds refer to and what their course of action might be.

The Comparison Study

From the propositions in the introduction, we argue that the interpretation of the sound event will be dependent upon the context within which it is presented. The contextual factors include the environmental setting and the experience of the person perceiving the sound. On this basis we have developed two main experimental hypotheses to test the theory of auditory affordances: (A) the function of representative, environmental sounds will be identified more readily than abstract, non-representative, sounds, (B) the interpretation of the function of a sound will be dependent upon the experience of the individual. These hypotheses were tested in an ITU at Southampton General Hospital. The ITU is an equipment-intensive environment, where individual patients may be attached to various pieces of monitoring equipment. Such equipment includes the following:

- humidifier (warms air before it enters the ventilator),
- infant warmer (warms infant from above),
- infusion pump (renal dialysis machine),
- ECG monitor (monitors heartrate),
- nutrition pump (feeds patient),
- pulse oxymeter (monitors pulse and oxygen levels),
- syringe pump (infuses drugs and blood into patient),
- ventilator (helps patient breathing),
- blood pressure monitor (continually monitors patient blood pressure).

Each of these monitoring facilities has alarm thresholds. For example, the ECG monitor typically has high and low thresholds (set individually for the patient). It may also have an alarm for no signal (if heart rate ceases) and an alarm to indicate a weak signal (due to disconnection or poor contact of one or more of the leads). There may be several infusion pumps, attached to a patient, which alarm when the infusion bag is nearly empty. The blood pressure monitor will also have high and low thresholds. In addition, each patient will have a 'crash' alarm button by their bed which is used by the nursing staff to call the 'crash team' if a patient has an arrest. This provides some idea of the range of different kinds of alarm sounds in the ITU. A single piece of equipment may possess up to 10 different auditory warnings.

From interviews with nursing staff, we established the need for the development of new alarm sounds. They reported problems with existing sounds, that is, alarms are typically:

- too loud,
- too high-pitched,
- too grating/irritating,
- too persistent,
- inappropriate,
- confusing (for example, ventilator alarm sounds like crash alarm).

These problems have been well reported in the literature (for example, Edworthy et al., 1991; Hass and Casali, 1995). For the purposes of this study, we were concerned with representational sounds that would inform the nurses as to which piece of equipment warranted their attention, rather than the specific nature of the problem, which is to be the subject of a future study.

As this was the first study of its kind, we focused primarily upon testing the theory of auditory affordances rather than on the proposed design methodology. The following account is a brief description of the study and its findings.

Method

The experimental method employed for the study was as follows.

Design   A standard 2 x 2 design was adopted, as illustrated in Figure 6.3. The sounds, both old and new, were presented to both groups, ITU (experts) and non-ITU staff (novices). The old sounds were sampled from the existing equipment, whereas the new sounds were taken from CD-ROM discs of sounds. They were not generated and selected by the ITU nurses in this instance, rather they were selected by two medical students who had experience of working in the ITU. The sounds were categorised as follows: *nomic* (for example, heartbeat for ECG monitor), *symbolic* (for example, nursery chime for infant warmer) and *metaphoric* (for example, bubbles for syringe pump). We recognise that the selection of sounds was purely subjective, but we can test the assumptions implicit in selecting the sounds in this experiment.

Equipment   Sounds were presented on a tape player and responses were recorded by pencil and paper.

Procedure   The old sounds were played first, followed by the new

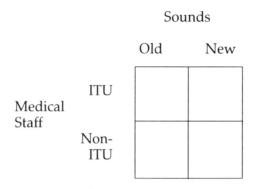

Figure 6.3   Experimental design

sounds. Only one trial was run. Each sound was played in turn to the medical staff who then had to denote the piece of equipment to which the sound referred (chosen from a list, see p. 121). Then they rated the appropriateness and urgency of the sound on the appropriate scales.

Participants The participants were 32 medical staff from Southampton General Hospital. Of these, 16 were experienced in ITU (experts) and 16 were general medical staff with non-ITU experience (novices). The two groups were matched for sex, age and experience.

Results

A Wilcoxon signed-rank test was used to analyse the results. The results showed that the ITU staff were significantly better at identifying the old sounds than at identifying the new (Z corrected for ties = -3.742, $p < 0.0005$), whereas the non-ITU staff were significantly better at identifying the new sounds than at identifying the old sounds (Z corrected for ties = -3.943, $p < 0.0001$). This is illustrated in Figure 6.4.

Analysis of the appropriateness rating scales was also undertaken. This is illustrated in Figure 6.5, which shows that the old sounds were rated as more appropriate for the function of warning than the new sounds, both by the experts (Z corrected for ties = 3.241, $p < 0.001$) and by the novices (Z corrected for ties = 2.821, $p < 0.005$).

Discussion

Although this investigation can only be considered a pilot study, and certainly not a test of the methodology itself, a number of interesting issues emerge. The first is that a more or less arbitrarily selected set of representational sounds (the 'new' sounds) are more readily associated with their designated referents than the traditional, abstract sounds actually used (the 'old' sounds). This is demonstrated by the fact that novices (those members of the nursing staff unused to the traditional alarms used in the ITU) were better able to make the appropriate mappings between the referents and the warning sounds for the new, than for the old, sounds. However the responses of the ITU staff (who already knew the meaning of many of the old sounds) showed that learned mappings are not easy to override in the face of potentially more appropriate sounds, as they matched the old, learned, set more accurately than the new. Thus these data show that, in the absence of already learned associations, more representational sounds may be intuitively easier to learn.

The results of the appropriateness ranking test shows an

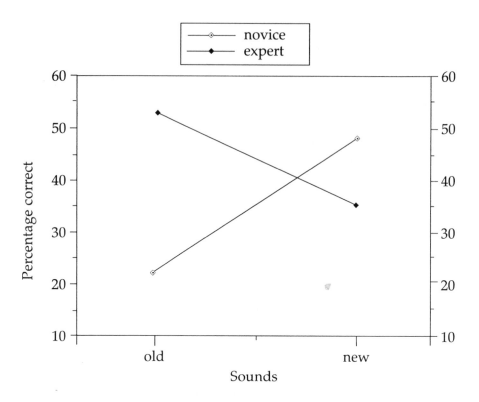

Figure 6.4   Recognition of equipment for old and new sounds

interesting paradox, however. Both sets of participants rated the new sounds as being less appropriate to the referents than the old sounds. This can be taken as a general resistance to radical departures in alarm design practice. In the case of the experienced ITU staff, this is probably a direct reflection of their already learned associations with old sounds, and is therefore easy to explain. On the other hand, the responses of the inexperienced staff suggest that they feel that the representational sounds (or at least the subset that was tested in this experiment) were less appropriate to the referents than the traditional sounds, even though an objective test of their ability to match them to the referents showed a superiority for the new sounds. This then raises the issue of whether people's ability to recognise sounds (as measured in the recognition tests) bears any relationship to their aesthetic responses to those sounds. In practice, whether or not a sound is suitable should surely be outweighed by its ability to be identified for a particular alarm function. The results of the first part of the experiment show that, for novices, who have no direct experience of ITU alarms, representational alarms might be more

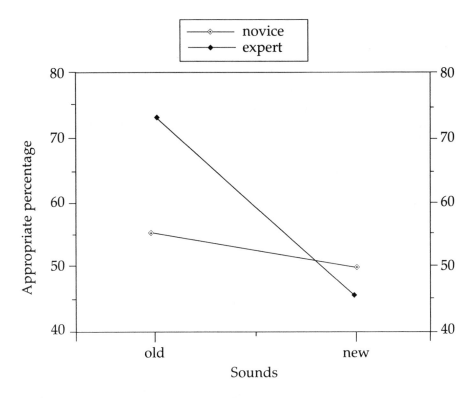

Figure 6.5 Appropriateness ratings for old and new sounds

effective if the primary aim of those alarms is to uniquely identify their referents.

Further one could conceive the notion of *learnt* affordances, where experience may lead the individual to learn and unlearn a particular representation. For example, ITU staff no longer appear to associate the 'heartbeat' sound as representative of the cardiovascular function, whereas for general staff it is. The ITU staff have learnt to associate other sounds with the cardiovascular system. This is consistent with the notion of learnt affordances. However, because the sounds are abstract, even after continual exposure over many years, performance is still very poor (see Figure 6.4). In fact, on a single trial of the new sounds, the non-ITU staff performed almost as well as the ITU staff with the old sounds, to which they have been exposed over many years.

In terms of the hypotheses (A and B) introduced earlier, we find support for hypothesis A for the non-ITU population but evidence against it for the ITU population. Hypothesis B is supported by virtue of the differences in the two populations' performance: ITU staff were significantly better at identifying the equipment with the old warnings whereas the non-ITU staff were significantly better at identifying the equipment with the new warnings. These findings lead us to two propositions. First, end-users must be involved in the design of the warning sounds. It is not acceptable to design sounds on their behalf, as we have shown in this study. Second, the theory of auditory affordances proposes that people understand sounds in the world in terms of their *potential for action*. This means that we should consider what the nurses do when they respond to these alarms, not what the alarms equipment is monitoring. This is a subtle, but important, shift in emphasis in our design methodology: the referents are the actions required by the nurses, for example to change a syringe pump or to resuscitate the patient.

Further research is required to investigate performance with the representational sounds over time. It is hypothesised that performance on the new sounds will improve with more trials and will be significantly superior than performance on the old sounds as exposure increases. It is also hypothesised that greater exposure to the new sounds would lead the ITU staff to relearn the original affordances.

General Conclusions

The results of the study are encouraging. People have problems with auditory warnings, and performance is poor with abstract sounds, particularly initial performance. Abstract sounds require the investment of considerable effort in learning, whereas representational sounds are more intuitive (at least to those who have not already invested effort in learning the abstract sounds). The experiment reported found an interaction between *experience* and *types of sound*. This requires further investigation, particularly into the nature of learned affordances and the development of the theory. It is likely, as suggested by the proposed methodology, that new warning sets will consist of both traditional and representational warnings. The combined strength of the best, and most appropriate, of both types of warnings should improve learning and identification performance for auditory warnings.

Acknowledgments

This chapter is heavily based upon two previous papers published by the authors in *Ergonomics* and *Applied Ergonomics* (see reference list). The authors are grateful to the publishers for being allowed to reproduce much of the material in this chapter.

References

Edworthy, J. and Stanton, N.A. (1995) 'A user-centred approach to the design and evaluation of auditory warnings: 1. Methodology', *Ergonomics*, **38**(11), 2262–80.

Edworthy, J., Loxley, S. and Dennis, I. (1991) 'Improving auditory warning design: relationship between warning sound parameters and perceived urgency', *Human Factors*, **33**, 205–31.

Gaver, W.W. (1989) 'The SonicFinder: an interface that uses auditory icons', *Human–Computer Interaction*, **4**, 67–94.

Gibson, J.J. (1979) *The Ecological Approach to Visual Perception*, New York: Houghton Mifflin.

Haas, E.C. and Casali, J.G. (1995) 'Perceived urgency of and response time to multi-tone and frequency-modulated warning signals in broadband noise', *Ergonomics*, **38** (11), 2281–99.

ISO/DIS 7001 *Specification for Public Symbols*, Geneva: International Organization for Standardization.

Lower, M.C., Patterson, R.D., Rood, G.M., Edworthy, J., Shailer, M.J., Milroy, R., Chillery, J. and Wheeler, P.D. (1986) 'The design and production of auditory warnings for helicopters. 1. The Sea King', Institute of Sound and Vibration Report No. AC527A.

Patterson, R.D. (1982) 'Guidelines for Auditory Warning Systems on Civil Aircraft', CAA paper 82017, London: Civil Aviation Authority.

Patterson, R.D., Edworthy, J., Shailer, M.J., Lower, M.C. and Wheeler, P.D. (1986) 'Alarm sounds for medical equipment in intensive care areas and operating theatres', Institute of Sound and Vibration Report No. AC589.

Stanford, L.M., McIntyre, J.W.R., Nelson, T.M. and Hogan, J.T. (1988) 'Affective responses to commercial and experimental auditory alarm signals for anaesthesia delivery and physiological monitoring equipment', *International Journal of Clinical Monitoring and Computing*, **5**, 111–18.

Stanton, N.A. (1993) 'Alarms in a coronary care unit', in E.J. Lovesey (ed.), *Contemporary Ergonomics*, London: Taylor & Francis.

Stanton, N.A. and Edworthy, J. (1994) 'Towards a methodology for designing representational auditory alarm displays', in S.A. Robertson (ed.), *Contemporary Ergonomics*, London: Taylor & Francis.

Stanton, N.A. and Edworthy, J. (1998) 'Auditory affordances in the Intensive Treatment Unit', *Applied Ergonomics*, **29**(5), 389–94.

7 The Perceived Urgency and Detection Time of Multitone Auditory Signals

ELLEN C. HAAS, *Aberdeen Proving Ground, Maryland, USA* and **JUDY EDWORTHY**, *University of Plymouth*

Introduction

The perceived urgency and detectability of auditory warning signals are important safety considerations. When designed correctly, auditory warning signals can improve operator performance and reduce accidents (Edworthy *et al.*, 1991), while inadequate signal design can cause problems. In environments in which there is a serious mismatch between the perceived (psychoacoustic) urgency of a warning and its situational urgency (the urgency associated with the state or condition that the signal represents), the listener may not perceive the urgency of the situation that the signal is trying to communicate (Edworthy *et al.*, 1991). When auditory warnings are not detectable within environments such as noisy workplaces, accidents may occur because a warning signal is either not heard (Wilkins & Acton, 1982), or heard and not heeded, such as in the case where the signal is sounded frequently and does not connote danger or urgency. In other instances, auditory warnings are so loud and distracting that people turn them off rather than use them (Patterson, 1982), as with highly repetitive auditory warnings on rapid-cycle machines such as printing presses and reversing alarms in industrial vehicles.

Patterson (1982) proposed a warning signal design methodology to help minimise these problems. His design specified a brief pulse of sound as a basic 'building block' for the signal. A pulse is defined as a sound with an onset, an offset (decay time) and a specific duration which is contained within one amplitude envelope. The pulse could be repeated several times, with intervals of silence between pulses. The resultant unit, referred to as a burst of sound, forms the basis of a complete warning sound. Edworthy *et al.* (1991) found that a wide variety of acoustic pulse and burst parameters had clear and consistent effects on the perceived urgency of auditory warnings and that subjects showed a high level of agreement about the perceived urgency of such warnings. They found that pulses with a higher fundamental frequency were rated as more urgent and that pulses containing a slow onset provided the highest ratings of perceived urgency. Finally, increasing the number of bursts in a warning increased signal urgency but also made the warning longer and gave it greater potential for becoming irritating and insistent. The work of both Patterson and Edworthy *et al.* was expanded in the study presented here.

The effects of wide ranges of specific signal parameters such as pulse fundamental frequency, pulse level and inter-pulse interval on both perceived urgency and detection (operationally defined as objective response time to auditory signals) have not been fully researched. The perceived urgency of and reaction time to signals with certain fundamental frequencies and sound pressure levels should be explored. Varying these parameters can improve resistance to masking and also may be applied to make the signal sound more urgent. Furthermore inter-pulse interval is another parameter, easily quantified and manipulated in electronic signals, which may influence the perceived urgency of warning signals. Although pulse level has been shown to affect both perceived urgency and response time (Edworthy *et al.*, 1991, Haas and Casali, 1995), the manner and extent of this relationship has not been well defined. These issues need to be addressed so that a model can be built for selecting auditory warning signals.

Research Objectives

The purpose of the present study was to determine the extent to which different auditory signal parameters affect the perceived urgency and response time associated with the perception of that signal. The objectives of this study were threefold: (1) to investigate the effect of pulse fundamental frequency, pulse level and inter-pulse interval on the perceived urgency of warning signals; (2) to investigate the effect of these variables on the response time to warning signals; and (3) to examine the relationship between perceived urgency and response time to warning signals. The research goal is to describe the relationship of certain parameters of auditory warning signals to a listener's perceived urgency and response time.

The major hypothesis of this experiment was that perceived urgency and subject response time would be affected by one or more of the experimental variables. Higher pulse level and fundamental frequency may result in greater perceived urgency of the signal and shorter response time. Inter-pulse interval may affect perceived urgency only. Shorter inter-pulse intervals may result in decreases in perceived urgency, and response time may remain unchanged.

The major goal of this research was to define parameters providing a quantifiable level of perceived urgency and response time. Knowledge of these parameters would provide a detailed, testable and functional description of the relationship among auditory warning parameters, perceived urgency and response time. This research will facilitate the design of future warning signals so that the most urgent sounds have the greatest perceived urgency and the shortest response time.

Method

A total of 30 subjects (15 males and 15 females) between the ages of 18 and 45 participated as paid subjects in this study. Subjects were civilian students attending Harford Community College, ranging in age from 22 to 45 years, with a mean age of 27.5 years. Subjects had 'unimpaired hearing', defined as an audiometric hearing threshold level (HL) of less than or equal to 15dBHL in each ear (Nicolosi *et al.*, 1989) at pure tone frequencies of 500 to 8000Hz. This bandwidth of hearing encompassed the frequency range of the signals used as stimuli. These qualifications were verified using pure-tone audiometry.

Test Facility and Apparatus

The test facility was housed at the US Army Research Laboratory at Aberdeen Proving Ground, Maryland, USA. An Industrial Acoustics Corporation (IAC) 1200-A audiometric booth was used for audiometric screening. Audiograms were administered using a Teledyne Avionics Model TA-20 Automatic Audiometer with Telephonics TDH-50P headphones.

An IAC 400-A Series semi-reverberant chamber, inside dimensions of 4.67m x 3.45m x 2.49m, was used as the test booth for all warning signal testing. The booth's ambient (no noise) Leq level measured at the centre of the subject's head was no more than 36dB (linear) at any third-octave band between 50 and 8000Hz. A Realistic Minimus-7 loudspeaker was positioned at the ear height (1.20m) of a seated 50th percentile subject (US Department of Defense, 1989), at the right-rear corner of the chamber, frontally incident to the centre of the chamber where the subject was seated. The loudspeaker emission and room acoustics provided a near-diffuse sound field (a volume in which the sound pressure level will assume a nearly constant value by +/− 1.0dB) in the vicinity of the subject's head. The subject was seated within this field, 2.82m from the speaker, in the centre of the chamber facing away from the loudspeaker. A blue dot positioned on the chamber wall served as a reference for helping the listener to maintain the head in a fixed position.

Warning signals were generated using Tucker Davis (TDT) SigGen (signal generation) software. A TDT System II with digital signal-processing capabilities interfaced with an IBM-PS2 series computer was used to store and play back the signals. In the response time task, the subject responded to the signals via a push button held in their dominant hand, from which the response latencies were obtained via computer clock. Signal calibration was performed daily using a Norwegian-Electronics Type 830 Real-Time Analyzer, with a Bruel and Kjaer Model 4165 one-half inch microphone. An intercom system consisting of microphones and loudspeakers in the test chamber and in the control room allowed the experimenter and subject to communicate with each other.

Experimental Design

A 3 x 3 x 3, full-factorial, within-subjects (repeated measures) design was used for data collection and to structure the primary data analysis. Subjects were treated as a random-effects variable and the remaining variables were treated as fixed-effects variables.

Independent variables   The three independent variables were pulse fundamental frequency, inter-pulse interval and pulse level. Pulse fundamental frequency was the fundamental frequency of the signals and included the first four harmonics of that frequency. The three fundamental frequencies used were 200, 500 and 800Hz. The signal with a fundamental frequency of 200Hz had components of 200, 400, 600, 800 and 1000 Hz. The signal with a fundamental frequency of 500Hz had components of 500, 1000, 1500, 2000 and 2500Hz. The signal with a fundamental frequency of 800Hz had components of 800, 1600, 2400, 3200 and 4000Hz. Signals with a fundamental frequency of 1000Hz and greater were avoided because the high-frequency components tend to make the signals sound aversive (Edworthy *et al.*, 1991).

Inter-pulse interval was the time elapsed from the end of the offset (decay) of one pulse to the onset of the next. The times were 0, 250 and 500ms. Pulse onset was the time from the waveform's increase from zero amplitude until it reached maximum output. Pulse offset was the time during which the pulse output fell from maximum amplitude to zero. The 0ms pulse interval contained no inter-pulse interval, because the end of the offset of one pulse was immediately followed by the onset of the next.

Pulse level was the rms (root mean square) sound pressure level (SPL) of the pulse (in dB LIN) at the centre of the subject's head, measured with the Norwegian Electronics real-time analyser in slow detector mode. The pulse levels were set at 5, 25 and 40dB LIN SPL above the ambient noise level of the chamber (a Leq of 36dB LIN SPL). Pulse levels in dB(A) and dB(C) were also recorded to permit comparison with data from other experiments.

The stimuli were 27 auditory signals, each signal consisting of a train of four pulses. Each pulse had a duration of 350ms, including an onset time of 25ms and an offset of 25ms. Stimuli with longer inter-pulse intervals had a longer duration. The total signal time was a minimum of 1.4s (at 0ms inter-pulse interval) and a maximum of 3.4s (at 500ms inter-pulse interval). Signals was presented in a randomised order across subjects and across trials.

Dependent variables   The two dependent measures were the free-modulus magnitude estimation rating of signal urgency, and subject response time in milliseconds to the warning signals.

Procedure

Each participant experienced four sessions, all in a single visit to the laboratory. In the first session, the subject received a pure-tone audiogram. If the subject fulfilled the hearing criteria described previously, he or she proceeded to the test chamber for the two remaining experimental sessions. If the subject did not fulfil the hearing criteria, they were advised of the audiogram results and released from the study.

In session two, the subject performed the free-modulus magnitude estimation task, in which he or she verbally rated the urgency of each auditory signal by assigning its urgency any number greater than zero. Within the randomised order of signal presentations, each signal was presented twice. Therefore the subject provided two magnitude estimation ratings for each of the 27 auditory signals.

After a 10-minute break, the subject performed the response time task (session three). The subject was informed that he or she would hear an auditory signal occurring at random times. The subject was instructed to respond to the auditory signal as soon as the signal sounded, by pressing a push button with his or her dominant hand. The subject performed 54 response time tasks (two for each signal).

After session three was completed, the subject again received a pure-tone audiogram (session four) to document that no shift in hearing threshold occurred as a result of participation in this test.

The order of the second and third experimental sessions was not counterbalanced. The magnitude estimation task was presented prior to the response time task to prevent subjects from having prior exposure to the signal stimuli (Stevens, 1971). Elimination of prior exposure is important in the magnitude estimation task.

Results

Data Reduction, Transformation and Correlations

Magnitude estimation data   To compare magnitude estimation judgments among subjects who responded using their own choice of number ranges in the free modulus procedure, each raw score was averaged and transformed, using a computational procedure originated by Lane *et al.* (1961) and described in detail in Haas (1993). In effect the transformation removed inter-subject variability due to different choices of moduli. The transformed data points, which comprised the arithmetic mean of the two responses for each signal per subject, were then analysed in a multivariate analysis of variance (MANOVA) for repeated measures.

Response time data   For each warning signal condition, the two response times provided by each subject were arithmetically averaged to yield one observation for analysis. Therefore 27 reaction time values per subject constituted the data set which was subjected to a MANOVA for repeated measures.

Correlations among dependent measures   The Pearson Product–Moment correlation between the magnitude estimation and response time data was small in magnitude ($r = 0.200$) but statistically significant ($p < 0.001$). The large sample size ($N = 783$) gave high statistical power to the test. Because the magnitude of the correlation was very small, the dependent variables were applied to separate MANOVAs for repeated measures to avoid potential reduction in statistical power.

Statistical Analysis: Magnitude Estimation Data

The MANOVA on the magnitude estimation data computed Wilk's criterion (U) for each source of variance, then converted to approximate F as per Rao (1973). A Huynh-Feldt correction (Vasey and Thayer, 1987) was applied to protect against violation of the assumption of homogeneity of covariance in the within-subject tests. The complete MANOVA table is contained in Table 7.1. The significant effects are as follows: interactions of fundamental

Table 7.1   Magnitude estimation data, MANOVA summary table*

Source of approx. variance	dvq	dfH	dfE	U	F	p
Between-subjects						
Subjects (S)	1	29				
Within-subjects						
Fundamental Frequency (F)	1	2	28	0.486	14.824	<0.001
F x S						
Inter-pulse interval (I)	1	2	28	0.566	10.722	<0.001
I x S						
Pulse level (L)	1	2	28	0.104	121.177	<0.001
L x S						
F x I	1	4	26	0.525	5.886	0.002
F x I x S						
F x L	1	4	26	0.557	5.170	0.003
F x L x S						
I x L	1	4	26	0.671	3.193	0.029
I x L x S						
F x I x L	1	8	22	0.471	3.083	0.017
F x I x L x S						

Notes

*Denominators used for each source of variance in the U tests appear as the second term in each grouping in the table.

dv = number of dependent measures; dfH = degrees of freedom for treatment effect; dfE = degrees of freedom for error effect; U = Wilk's likelihood ratio statistic; p = significance of approximate F.

frequency, pulse level and inter-pulse interval (F (8,22) = 3.083, p = 0.017), fundamental frequency and inter-pulse interval (F (4,26) = 5.886, p = 0.002), fundamental frequency and pulse level (F (4,26) = 5.170, p = 0.003) and inter-pulse interval and pulse level (F (4,26) = 3.193, p = 0.029); main effects of fundamental frequency (F (2,28) = 14.824, p < 0.001), inter-pulse interval (F (2,28) = 10.722, p < 0.001) and pulse level (F (2,28) = 121.177, p < 0.001). These significant effects were explored on a post hoc basis by using a Newman-Keuls Sequential Range test performed at p < 0.05.

Fundamental frequency by inter-pulse interval by level interaction   The post hoc test results for the three-way interaction of fundamental frequency, inter-pulse interval and pulse level (Table 7.2) indicated that perceived urgency increased differentially as fundamental frequency and pulse level increased, and as inter-pulse interval decreased. Pulses with the highest fundamental frequency (800Hz) and sound pressure level above ambient (40dB LIN) and the shortest inter-pulse interval (0ms) were judged to be significantly more urgent than any other signals. Signals with higher fundamental frequencies (500 and 800Hz) at the highest level above ambient (40dB LIN) had a significantly greater perceived urgency than those with the 200Hz fundamental frequency at every inter-pulse interval. Pulses at the lowest sound pressure level above ambient (5dB LIN), at any combination of fundamental frequency and inter-pulse interval, were judged to be significantly less urgent than pulses at higher sound pressure levels (25 and 40dB LIN).

Fundamental frequency by inter-pulse interval interaction   The post hoc test results for the interaction of fundamental frequency and inter-pulse interval (Figure 7.1) indicate that pulse perceived urgency increases differentially as fundamental frequency increases and as inter-pulse interval decreases. Within all inter-pulse intervals, perceived urgency was seen to increase significantly from 200 to 500Hz, but there was no significant increase in perceived urgency from 500 to 800Hz. At 500 and 800Hz, all inter-pulse intervals were significantly different, with 0ms inter-pulse intervals judged to have the highest level of urgency, and 500ms inter-pulse intervals judged to have the least. At the 200Hz fundamental frequency, there was no significant difference in perceived urgency for signals with inter-pulse intervals of 0 and 250ms. Signals with 500ms inter-pulse intervals were judged to be significantly less urgent than all other inter-pulse intervals, at all fundamental frequencies. These results indicate a strong perceptual difference that is elicited by temporal differences in signal construction. It should also be noted that all inter-pulse intervals are ordinal, in that the relative positioning of

Table 7.2 Newman-Keuls test: magnitude estimation data, pulse format by inter-pulse interval by pulse level

Fundamental frequency	Inter-pulse interval (ms)	Pulse level (dB LIN)	Mean magnitude estimate	Significance*						
200	0	5	2.251	A						
200	0	25	5.531			C D				
200	0	40	9.961						G	
200	250	5	2.552	A						
200	250	25	5.088			C D				
200	250	40	9.444						G	
200	500	5	2.226	A						
200	500	25	4.876			C				
200	500	40	8.352					F		
500	0	5	3.359	A B						
500	0	25	7.826					E F		
500	0	40	11.876							H
500	250	5	2.740	A B						
500	250	25	5.949			C D				
500	250	40	11.291							H
500	500	5	2.652	A B						
500	500	25	4.944			C				
500	500	40	9.963						G	
800	0	5	3.732		B					
800	0	25	7.133					E		
800	0	40	12.780							I
800	250	5	2.932	A B						
800	250	25	6.137				D			
800	250	40	11.029							H
800	500	5	2.473	A						
800	500	25	5.145			C D				
800	500	40	10.943							H

*Levels of magnitude estimation with the same letters are not significantly different from each other at the 0.05 level.

each level of each independent variable does not change. For example, Figure 7.1 shows that the perceived urgency of signals with 0ms inter-pulse intervals is always greater than signals with 250ms inter-pulse intervals, which are greater than signals with 500ms inter-pulse intervals, across every fundamental frequency. Therefore both fundamental frequency and inter-pulse interval can be interpreted independently of this interaction.

Fundamental frequency by pulse level interaction   As indicated in Figure 7.2, increases in pulse level lead to greater perceived urgency at all fundamental frequencies. Perceived urgency ratings between each increase in pulse level differ by at least 2.5 scale points on a 12-point scale range shown. This indicates the strong perceptual difference that is elicited by level differences in signal construction, even when the same

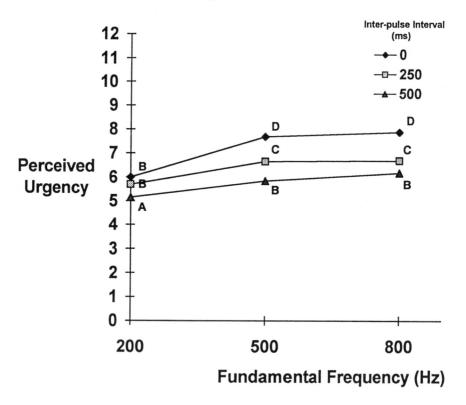

Means with different letters are significantly different at p < 0.05

Figure 7.1   Mean perceived urgency for the fundamental frequency x inter-pulse interval interaction

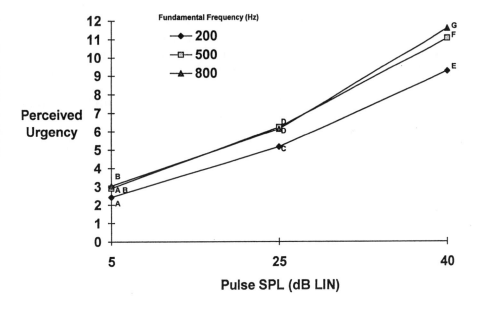

Means with different letters are significantly different at p < 0.05

Figure 7.2   Mean perceived urgency for the fundamental frequency x pulse SPL (dB LIN) interaction (each fundamental frequency plotted separately)

frequency components are used. Within pulse levels, some increases in fundamental frequency do not produce a significantly greater perception of urgency. At 5dB LIN SPL, pulses of 500Hz fundamental frequencies are not perceived as being significantly more urgent than those of 200Hz or significantly less urgent than those of 800Hz. At 25dB LIN SPL, pulses with a 500Hz fundamental frequency are not perceived as significantly more urgent than pulses with an 800Hz fundamental frequency. However, at 40dB LIN SPL, the higher the fundamental frequency, the significantly greater the perceived urgency of the signal. This interaction between fundamental frequency and pulse level was analysed to determine whether it would preclude the interpretation of the fundamental frequency main effect. As seen in Figure 7.2, only 200Hz signals may be interpreted independently of this interaction at all pulse levels, because each level is ordinal with respect to the others. The 500Hz and 800Hz are not ordinal with respect to each other across all levels. Therefore this interaction allows the unambiguous interpretation of the 200Hz fundamental frequency, and caution is used in interpreting main effects for 500 and 800Hz fundamental frequencies.

This interaction was also analysed to determine whether it would preclude the interpretation of the main effect of pulse level. As can be seen in Figure 7.3, all pulse levels are ordinal with respect to all others. This particular interaction allows the unambiguous interpretation of the main effect of pulse level.

Inter-pulse interval and pulse level interaction   Figure 7.4 shows that the perceived urgency of the three inter-pulse intervals increases in slightly differential fashion with an increase in pulse level. At 40dB

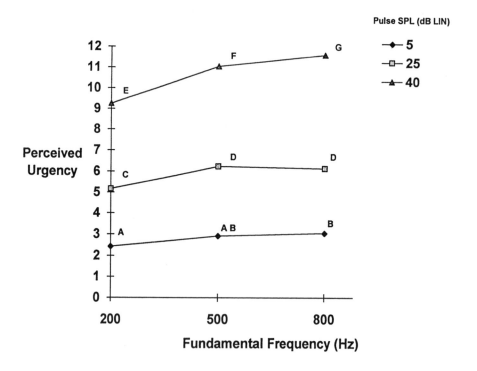

Means with different letters are significantly different at $\underline{p} < 0.05$

Figure 7.3   Mean   perceived   urgency   for   the   fundamental frequency **x** pulse SPL (dB LIN) interaction (each SPL plotted separately)

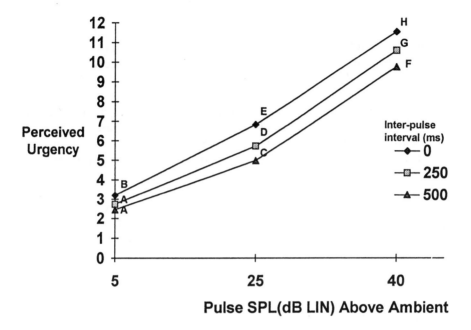

Means with different letters are significantly different at p < 0.05

Figure 7.4 **Mean perceived urgency for the inter-pulse interval x pulse SPL interaction**

LIN SPL, the perceived urgency of the 0ms inter-pulse interval is significantly greater than that of the 250ms interval, which in turn is significantly greater than that of the 500ms interval. At 25dB SPL, the same significant differences repeat. At 5dB SPL, pulses with 250 and 500ms inter-pulse intervals are not perceived as being significantly different. As pulse level increases, the 0ms inter-pulse interval shows the greatest increase in perceived urgency. This interaction allows the independent interpretation of the inter-pulse interval and the pulse level main effects because all inter-pulse intervals and pulse levels are ordinal with respect to the others.

Main effects Significant main effects involving the signal variables were further analysed with the Newman-Keuls procedures, and mean magnitude estimation values and standard deviations for each level are shown in parentheses (mean; sd) in the ensuing text. The interaction data permit the conclusion that subjects rated signals with a fundamental frequency of 200Hz (mean magnitude estimate of 5.606, sd of 3.413) as sounding significantly less urgent than those containing 500Hz (6.720, 4.064) or 800Hz (6.984; 4.371) fundamental frequencies. There is no significant difference in perceived urgency between the 500Hz and 800Hz signals, except when the signals are presented at the highest sound pressure level (40dB LIN), where 800Hz signals are judged to be significantly more urgent than those of 500Hz. As for the main effect of inter-pulse interval, all means were significantly different from each other. Subjects rated signals with an inter-pulse interval of 0ms as sounding significantly more urgent (7.211; 4.284) than those containing intervals of 250ms (6.374; 3.941). Signals containing intervals of 500ms (5.724; 3.654) sounded significantly less urgent than the other inter-pulse intervals. All mean signal levels were significantly different from each other. The means and standard deviations for the magnitude estimation of the pulse level main effect indicate that subjects rated signals 40dB LIN above ambient (10.652; 3.247) signals as significantly more urgent than those at 25dB LIN above ambient (5.866, 2.040). Signals at 5dB LIN (2.801; 1.484) above ambient were perceived as being significantly less urgent than the other two levels.

Response surface procedures (Myers, 1976) were performed for magnitude estimation ratings to determine what values of the independent variables (fundamental frequency, level and inter-pulse interval) were optimum (greatest) for predicting perceived urgency. The stepwise regression equation included second-order effects, and interactions were included for those variables known to interact. The regression is significant ($F = 192.303$, $p < 0.001$). The beta values for fundamental frequency, fundamental frequency times level, fundamental frequency*2 (fundamental frequency squared), and level*2 (level squared) are significant ($p > 0.05$). The remaining values are not significant ($p < 0.05$), but were included in the regression equation because they are variables of interest. The equation indicated that perceived urgency is greatest when pulse level is highest (40dB LIN above ambient), when fundamental frequency is greatest (800Hz), and when inter-pulse interval is smallest (0ms between pulses).

Stastical Analysis: Response Time Data

The response time data for one male subject were eliminated because he did not respond properly to the reaction time task. Therefore the response time data includes 29 subjects, rather than 30. The MANOVA on the magnitude estimation data computed Wilk's criterion (U) for each source of variance, then converted to approximate F as per Rao (1973). A Huynh-Feldt correction (Vasey and Thayer, 1987) was applied to protect against violation of the assumption of homogeneity of covariance in the within-subject tests. The complete MANOVA table can be seen in Table 7.3. The significant

Table 7.3   Response time data, MANOVA summary table*

Source of approx. variance	dv	dfH	dfE	U	F	p
Between-subjects						
Subjects (S)	1	28				
Within-subjects						
Fundamental Frequency (F) F x S	1	2	27	0.751	4.482	0.021
Inter-pulse interval (I) I x S	1	2	27	0.868	2.051	0.148
Pulse level (L) L x S	1	2	27	0.220	47.900	<0.000
F x I F x I x S	1	4	25	0.743	2.161	0.103
F x L F x L x S	1	4	25	0.838	1.210	0.332
I x L I x L x S	1	4	25	0.906	0.640	0.634
F x I x L F x I x L x S	1	8	21	0.414	3.721	0.007

Notes

*Denominators used for each source of variance in the U tests appear as the second term in each grouping in the table.

dv = number of dependent measures; dfH = degrees of freedom for treatment effect; dfE = degrees of freedom for error effect; U = Wilk's likelihood ratio statistic; p = significance of approximate F.

effects are as follows: interactions of fundamental frequency, pulse level and inter-pulse interval (F (8,21) = 3.721, p = 0.007), and the main effects of fundamental frequency (F (2,27) = 4.482, p = 0.021) and pulse level (F (2,27) = 47.900, p = 0.000). These significant effects were explored, on a post hoc basis, by using a Newman-Keuls Sequential Range test performed at p < 0.05.

Fundamental frequency by inter-pulse interval by pulse level interaction
The post hoc test results for the interaction of fundamental frequency, inter-pulse interval and pulse level (Table 7.4) indicate that response time to signals decreases differentially as fundamental frequency and pulse level increase, and as inter-pulse interval decreases. Pulses with the lowest fundamental frequency (200Hz), level above ambient (5dB LIN), and shortest inter-pulse interval (0ms) had a response time that was significantly greater than most of the other pulse parameter combinations. The shortest response time was obtained for signals with 500Hz, at 0ms between pulses and 40dB LIN. However this response time was not significantly shorter than for most of the other pulses. At the level of this three-way interaction, most pulse response times were not significantly different from one another.

Main effects   Significant main effects involving the signal variables were further analysed with the Newman-Keuls procedures. Mean response times and standard deviations for each level are shown in parentheses (mean; sd) in the ensuing text.
    The interaction data permit the conclusion that signals with a fundamental frequency of 800Hz (mean response time of 321ms, sd of 98) had a significantly shorter response time than those containing 200Hz (340; 84). There is no significant difference in response time between signals of 500Hz (329; 88) and 800Hz. All pulse level means were significantly different from each other. Response times for signals 40dB LIN above ambient (303; 80) were the shortest, followed by signals 25dB above ambient (324; 84) and those at 5dB LIN above ambient (363; 96).

Response surface procedures   Response surface procedures (Myers, 1976) were performed for subject response times to determine what values of the independent variables (fundamental frequency, level and inter-pulse interval) were optimum (greatest) for each dependent variable. The stepwise regression equation included second-order effects, and interactions were included for those variables known to interact. The regression is significant (F = 184.20, p < 0.001). The beta values for fundamental frequency, fundamental frequency times time, fundamental frequency*2, and level*2 are significant. The remaining values are not significant (p < 0.05), but were included in

Table 7.4 **Newman-Keuls test: response time data, pulse format by inter-pulse interval by pulse level**

Fundamental frequency	Pulse interval (ms)	Inter-pulse level (dB LIN)	Mean Response Time (ms)	Significance*
200	0	5	409	E
200	0	25	327	A B C D
200	0	40	326	A B C D
200	250	5	350	B C D
200	250	25	339	A B C D
200	250	40	309	A B C
200	500	5	367	C D E
200	500	25	318	A B C D
200	500	40	316	A B C D
500	0	5	368	C D E
500	0	25	353	B C D
500	0	40	286	A
500	250	5	370	D E
500	250	25	323	A B C D
500	250	40	315	A B C D
500	500	5	338	A B C D
500	500	25	316	A B C D
500	500	40	296	A B
800	0	5	342	A B C D
800	0	25	296	A B
800	0	40	300	A B
800	250	5	372	D E
800	250	25	325	A B C D
800	250	40	289	A
800	500	5	353	B C D
800	500	25	320	A B C D
800	500	40	296	A B

*Levels of response time with the same letters are not significantly different from each other at the 0.05 level.

the regression equation because they are variables of interest. The equation indicated that the response time of signals is greatest when pulse level is highest (40dB LIN above ambient), when fundamental frequency is greatest (800Hz) and when inter-pulse interval is smallest (0ms between pulses).

Discussion

This experiment was successful in providing a detailed, testable description of the dependence of perceived urgency and listener response time on specific frequency, amplitude and temporal parameters of auditory warning signals. As in any controlled laboratory study or experiment, the number of parameters was limited, but it is important to note that each selected parameter yielded significant effects on a majority of the urgency and response time metrics. Therefore the results of this study may be used to govern the design of future related experiments and to provide some insight into the design of auditory signals which must convey a sense of urgency. The ensuing discussion will discuss both subjective perceived urgency and objective response time.

Signal Perceived Urgency

The effect of pulse level on perceived urgency has been explored by few researchers previous to this study. Haas and Casali (1995) found that there is a strong relationship between signal level and perceived urgency. This experiment confirms this hypothesis, and expands it to apply to a greater range of sound pressure levels (5dB LIN to 40dB LIN) above ambient. The post hoc tests clearly indicate that the higher pulse levels resulted in significantly greater perceived urgency. The results of this study may be applied to the practical application of the design of signals within the recommended signal sound pressure level limits (15 to 30dB above ambient) which are specified within international design standards (ISO 1986) and research recommendations (Patterson, 1982).

Hellier and Edworthy (1989) and Haas and Casali (1995) found that the interval between signal pulses had a consistent effect on the perceived urgency of warning signals. Their results were replicated with different pulse interval durations in this study, which indicates that signals with shorter inter-pulse intervals were rated as significantly more urgent. Signals with no inter-pulse interval (0ms) were rated as the most urgent of all.

The effect of pulse fundamental frequency on perceived urgency had not been investigated at any great length prior to this study. The post hoc tests clearly indicate that, although the higher fundamental frequencies resulted in significantly greater signal perceived urgency, there was no difference in the perceived urgency of signals with 500 and 800Hz fundamentals, except at 40dB SPL, where signals of 800Hz were found to be more urgent. One topic for future research could involve defining the point at which fundamental frequency no longer produces differences in perceived urgency.

Signal Response Time

Signal response time was significantly influenced by both level and spectral factors. The difference in mean response times for signals 5 and 40dB LIN above ambient is 60ms. In terms of human response, this is an extremely short interval, but is of practical significance for certain applications. For instance, a response time difference of 60ms may result in crash avoidance for a pilot performing nap-of-earth manoeuvres in a fighter jet where fast response is critical. Although less significant, response time of this time interval may have some significant effect in the response time to a reversing alarm or an alarm emanating from medical equipment in an operating room or an intensive care setting. In some other applications, such as responding to a fire alarm, a difference of such small time magnitude may not make a practical difference. In any case, however, there appears to be little or no disadvantage in designing the signal to elicit the shortest response time.

Perceived Urgency and Response Time Relationships

As was found in Haas and Casali (1995), the correlational analysis for the magnitude estimation and response time data indicates that, as perceived urgency increases, response time to the signal decreases. The result was supported by the significance of the correlation between the metrics.

Pulse level and fundamental frequency are the signal design characteristics which consistently affected all measures of perceived urgency and response time. A significant difference between pulse levels and fundamental frequencies was detected in both the magnitude estimation and the response time measures. Across the levels used, the higher the pulse level, the greater the perceived urgency of the signal and the shorter the response time to the signal; the greater the fundamental frequency, the shorter the response time to the signal. Perceived urgency increased as fundamental frequency increased from 200 to 500Hz, but was not significantly greater at 800Hz, except at the highest sound pressure level (40dB LIN). An increase in fundamental frequency did not result in an increase in perceived urgency, except at relatively high sound pressure levels above ambient, but not at sound pressure levels previously described as being recommended for signal design (15–30dB above ambient) within ISO (1986) and Patterson (1982).

Application Issues

There are several military and industrial environments where the perceived urgency and response time to auditory warning signals should be of issue. This is especially true in environments which are highly dynamic and complex, and whose tasks involve high levels of workload. Military and civilian aircraft and helicopter cockpits use multiple acoustic signals, many of which can be activated simultaneously (Patterson, 1982). Short response time to cockpit signals is important to a pilot performing rapid tactical manoeuvres in a high-speed aircraft, where fast manual response is often essential. Hospital and operating room personnel (especially anaesthesiologists, as noted by Gaba and Howard, 1995) often assemble many pieces of patient physiological monitoring equipment which employ a wide variety of acoustic signals to signify change in patient condition and equipment operation status. Utility and chemical plant control rooms utilise multiple aural signals to warn operators of impending conditions and to orient attention towards specific control panel locations. In each of these and other applications, proper signal parameter specification and design can aid the listener in determining the need for rapidity of response to each signal.

References

Edworthy, J., Loxley, S. and Dennis, I. (1991) 'Improving auditory warning design: Relationship between warning sound parameters and perceived urgency', *Human Factors*, **33**, 205–31.

Gaba, D.M. and Howard, S.K. (1995) 'Situation Awareness in Anesthesiology', *Human Factors*, **37**, 20–31.

Haas, E. (1993) *The perceived urgency and detection time of multi-tone and frequency-modulated warning signals in broadband noise* (doctoral dissertation), Virginia Polytechnic Institute and State University, Virginia.

Haas, E. and Casali, J.G. (1995) 'Perceived urgency and response time to multi-tone and frequency-modulated warning signals in broadband noise', *Ergonomics*, **38**(11), 2281–99.

Hellier, E. and Edworthy, J. (1989) 'Quantifying the perceived urgency of auditory warnings', *Canadian Acoustics*, **17**, 3–11.

ISO (1986) *Danger Signals for Work Places – Auditory danger signals*, (ISO 7731-1986 (E)), Geneva: International Organization for Standardization.

Lane, H., Catania, A. and Stevens, S. (1961) 'Voice Level: Autophonic scale, perceived loudness and the effects of sidetone', *Journal of the Acoustical Society of America*, **33**, 160–67.

Myers, R.H. (1976) *Response Surface Methodology*, Ann Arbor, MI: Edwards Brothers.

Nicolosi, L., Harryman, E. and Kreshek, J. (1989) *Terminology of Communication Disorders: Speech–language–hearing*, Baltimore: Williams and Wilkins.

Patterson, R.D. (1982) 'Guidelines for Auditory Warning Systems on Civil Aircraft', CAA paper 82017, London: Civil Aviation Authority,.

Rao, C.R. (1973) *Linear Statistical Inference and its Applications*, (2nd edn), New York: John Wiley.

Stevens, S.S. (1971) 'Issues in psychophysical measurement', *Psychological Review*, **78**(5), 426–50.

US Department of Defense (1989) 'Military Standard 1474D', Human engineering design criteria for military systems, equipment and facilities, Washington DC: Department of Defense.

Vasey, M.W. and Thayer, J.F. (1987) 'The continuing problem of false positives in repeated measures ANOVA in psychophysiology: A multivariate solution', *Psychophysiology*, **24**, 479–86.

Wilkins, P.A. and Acton, W.I. (1982) 'Noise and accidents – A review', *Annals of Occupational Hygiene*, **2**, 249–60.

8 A Psychophysiological Evaluation of the Perceived Urgency of Auditory Warning Signals

JENNIFER L. BURT, *NASA Langley Research Center*, DEBBIE S. BARTOLOME-RULL, *Lockheed Martin Engineering and Sciences Company*, DANIEL W. BURDETTE, *Lockheed Martin Engineering and Sciences Company* and J. RAYMOND COMSTOCK, *NASA Langley Research Center*

Introduction

Human factors research in aviation is taking on new importance as the role of the flight crew changes from that of manipulators of aircraft controls to system monitors and decision makers. Technological advances have made aircraft safer by promoting the development of systems which are better designed and cockpit instrumentation which is more reliable, but these advances may have negative psychological effects on the flight crew. While new technology requires that pilots continuously expand their range of skills, the increasing level of cockpit automation may also result in reduced mental activity, boredom and, consequently, hazardous states of awareness (Pope and Bogart, 1992).

Issues currently under investigation in aviation psychology include cockpit communication, cockpit resource management, pilot judgment, pilot reliability, situation awareness and workload assessment. Of particular interest in the automated flight-deck and in other critical monitoring situations is the psychological construct of perceived or psychoacoustic urgency of auditory warnings. This is of interest in aviation psychology because a greater understanding of this phenomenon may lead to the design of safer, more effective and more reliable aircraft warning systems.

Cockpit caution and warning signals are presented to attract pilot attention and provide pertinent information about current or impending hazards. Along with communication with other members of the air and ground crew, the auditory channel is also used for the presentation of cautions, warnings and advisories. Examples of auditory warnings include horns, whistles, sirens, bells, buzzers, chimes, gongs, clackers, pure tones, synthesised voice messages and tones which vary in their pulse and burst characteristics. Although Irving (1981) found that pilots responded more quickly and accurately to voice warnings than to tone warnings, Wheale (1981, 1982, 1983) found that voice warnings actually produced slower reaction times than other types of auditory warnings. Wheale's findings, along with the complex implementation and time limitations of voice messages, have resulted in the continued use of tone warnings and the continued research and refinement of voice warnings.

The first auditory warnings were employed to alert aircrew to dire emergencies such as landing gear up, engine fire and power control failure. With the development of faster aircraft and the growing significance of response time, however, auditory warnings have been used to provide routine flight information such as airspeed, altitude, angle of attack, pitch and roll. Although most information is still presented visually through panel indicators, modern military aircraft incorporate many auditory signals as warnings because they have several specific advantages over visual signals.

Auditory warnings were originally implemented in aircraft because signal lights were considered to be ineffective, and auditory warnings conserved space on the already crowded panel of the cockpit. Continued research has shown that auditory warnings have many advantages other than just conserving space in the cockpit's limited area. They alert the pilot to dangerous or potentially dangerous conditions regardless of head position and direction of gaze, as well as provide sensory inputs which are less disrupted by anoxia and positive G-forces than are visual inputs (Munns, 1971; Doll et al., 1984; Doll and Folds, 1985; Edworthy et al., 1991). Auditory warnings also enable the pilot to fly head-up for longer periods of time and provide relief from the constant monitoring of visual warning displays. Since auditory warnings reduce the need to scan the instrument panel visually, the pilot's visual workload is decreased and the probability and speed with which one may react to emergency situations is increased (Bertone, 1982; Doll and Folds,

1985). Several studies conducted under simulated conditions with experienced pilots have shown that auditory warnings produce faster response times than the visual warnings presented on panel indicators (Reinecke, 1976, 1981; Wheale, 1981, 1982, 1983). In addition to advantages illustrated by improved performance, pilots have also reported preferences for auditory signals (Doll and Folds, 1985; Axelsson and Stoby, 1991).

While auditory warnings offer many benefits, they also create special problems. Many of the existing warnings are described as being too loud, insistent, startling and distracting; they disrupt thought and communication and are likely to be viewed by the pilot as annoying rather than helpful (Munns, 1971; Doll *et al.*, 1984; Doll and Folds, 1985; Folds, 1985; Wheale, 1983; Rood, 1989; Rood *et al.*, 1985; Thorning and Ablett, 1985; Edworthy *et al.*, 1991). Similar reactions to auditory warnings have been found in industrial settings and hospital intensive care units as well (Kerr, 1985; Kerr and Hayes, 1983; Lazarus and Höge, 1986; Meredith and Edworthy, 1994). Marshall (1987) conducted a survey of experienced pilots and described the typical reaction of aircrew to auditory warnings in the following statement:

> The way things are at the moment, a plethora of bells, buzzers, clackers and horns tend to alarm an operator – most of them make flight deck speech impossible, and the immediate reaction is to try to switch the damn thing off before addressing the problem associated with it.

Also, pilots may come to rely too heavily on the multitude of auditory warnings and become lulled into a false sense of security (Munns, 1971; Folds, 1985).

One of the most significant concerns that pilots have about auditory warnings is that they lack a sense of priority. Patterson (1982) found that pilots said of auditory warning signals: 'When a warning occurs, it is usually either a false warning or the direct result of a standard flight procedure. ... Even when a true warning occurs, it almost always indicates a potential problem rather than a sudden emergency.' Pilots would like the priority level of auditory warnings presented in the cockpit to be indicated so that they may immediately assess the urgency of the situation (Veitengruber *et al.*, 1977). For example, a warning signifying an emergency situation, such as an engine fire, should be designed so that it is perceived as being extremely urgent. By providing auditory warnings which are designed with sound parameters capable of matching a pilot's psychoacoustic perception of urgency, warning urgency level can be quickly discerned, and time and resources may be saved during critical phases of flight.

While perceptual theorists have researched the psychological quality of various signals throughout the century, it was not until recently that such investigative findings were applied in operational settings such as aircraft cockpits, hospital recovery rooms and nuclear power stations. The work of Edworthy and her colleagues, for example, provides some very pertinent and applicable psychoacoustic research. Edworthy *et al.* (1991) conducted an extensive series of studies which showed that fundamental frequency, harmonic series, amplitude envelope shape, delayed harmonics and temporal and melodic parameters such as speed, rhythm, pitch range and melodic structure all have clear and consistent effects on perceived urgency. According to Edworthy (1994): 'It is possible ... to convey an appropriate level of urgency through the warning itself by the way in which the acoustic features of the warning are manipulated.' By following the steps of an auditory warning construction outlined by Patterson (1982, 1989), Edworthy *et al.* (1991) were able to construct 13 auditory warnings which were consistently ranked from most urgent to least urgent by their subjects. Patterson's four steps include determining the appropriate level of loudness, designing a small pulse of sound, incorporating the pulse into a longer burst of sound, and forming a complete auditory warning using bursts of sound followed by short periods of silence. In their first series of experiments, Edworthy *et al.* (1991) found that subjects were able to rank consistently the perceived urgency of small bursts of sound varied along spectral and temporal parameters. Urgency mapping is therefore possible for small bursts of sound lasting approximately 100–300ms as well as for longer bursts of sound lasting approximately 2s (Edworthy *et al.*, 1991).

The ability of subjects to rank consistently the perceived urgency of small bursts of sound is relevant to the present experiment because only signals of a short duration will allow the recording of event-related brain potentials (ERPs). An ERP represents an average of a number of neurological reactions which are generated in response to specific and brief stimuli, such as a flash of light or a pulse of sound. Since ERPs may be recorded only in response to brief stimuli, constant temporal characteristics of the stimuli are necessary. Auditory warning ERP data were desired in the present investigation in an attempt to relate 'specific, ongoing changes in brain activity to discrete psychological states and events' (Andreassi, 1989).

The ability to measure the brain's response to auditory stimuli was also required in this investigation to examine the general level of arousal in response to different auditory warning signals. The electroencephalogram (EEG), or brainwave, represents the electrical activity of the brain measured from the surface of the scalp. Since the work of Berger (1929), EEG has been used clinically and experimentally as an unobtrusive psychophysiological technique for estimating arousal and its relation to attention and effort (Andreassi, 1989; Kramer, 1991).

Since a search of the literature revealed a paucity of research concerning physiological responses to auditory warning signals, the present study examined the relationship of auditory parameters to perceived urgency by measuring the physiological responses associated with EEG and ERP data as well as subjective assessments and reaction time. These physiological measures were chosen because '[f]requency analysis and the averaging of event-related potentials are powerful tools in the analysis of cognitive processes' (Empson, 1986). EEG frequency analysis involves determining the power, or amount of activity, in the theta (4–7Hz), alpha (8–12Hz) and beta (13–20Hz) frequency bands within an EEG segment. An increased amount of theta power indicates a state of diminished arousal; increased alpha power reflects relaxed wakefulness; and increased beta power corresponds with alert attentiveness (Lindsley, 1960). In addition to examining the absolute power of individual frequency bands, it is also possible to compare band power ratios such as fast to slow brainwave activity (beta/(alpha + theta)) (Pope *et al.*, 1994). These ratios are important in that they convey information concerning the differential activity among frequency bands.

The analysis of ERP amplitudes reveals waveform components which capture the intensity of the brain's response to stimulation, with larger amplitudes indicating a more intense response. In a study employing auditory-evoked responses to a voice warning, a buzzer and a pure tone, Utsuki *et al.* (1977) found that auditory-evoked responses differed, with the largest amplitude of response occurring in the case of pure (sinusoidal) tones. This brainwave evaluation technique was useful in the present experiment because it allowed the examination of the immediate perceptual response of arousal elicited by an auditory warning. The amplitudes of both the immediate or early components (<200ms) of the ERP which reflect the degree of perceptual processing and a later component (about 400ms) which reflects cognitive processing were examined. The present study evaluated the amplitude differences as a function of manipulating the sound parameters of an auditory warning and the importance of the warning to the task at hand. It was expected that the ERP response to the high urgency warning (perceived) would be characterised by increased amplitudes of the early and later components, reflecting an increase in arousal and attention.

Physiological measures used in this study can assist in determining the degree to which the subject's response is due to the sound parameters, the situational context or an interaction of the two. In the design of this experiment, two urgency manipulations were used as independent variables, and these are described below.

Perceived Urgency

Perceived urgency was defined as the inherent sense of urgency conveyed by the warning sound parameters. These sound parameters, which are described in Table 8.1, were selected so that one warning represented high urgency, one represented moderate urgency and one represented low urgency, based on their psychoacoustic characteristics. (Details of sound stimuli generation are presented in the 'Method' section below.)

The sound parameters for these three warnings were patterned on the research of Edworthy *et al.* (1991) which demonstrated that sounds have differences in perceived urgency which can be ranked. The temporal sound parameters of the auditory warnings which were manipulated in Edworthy *et al.* (1991) were, however, held constant owing to stimuli constraints for the collection of ERPs. The time constraints imposed by the recording of ERPs precluded the use of manipulations, such as slow or delayed onset and offset of the amplitude envelope and delayed onset of the higher harmonics.

Situational Urgency

Situational urgency was operationally defined as the probability that control of the tracking target would be lost. A high level of situational urgency (high probability of the target control being lost) should demand a greater allocation of attention and resources by the operator than does a moderate or low level of situational urgency. By varying situational urgency, it is possible to determine if a subject's response to a warning is altered as a result of the situation signalled by the warning. For a given subject, each of the three warnings was randomly assigned a single proportion of tracking task failure, as shown in Table 8.2.

The attentional engagement variable was manipulated through the use of manual and automated conditions of the tracking task. The manual tracking condition required continuous performance of

Table 8.1   Sound parameters

Fundamental frequency (Hz)	Harmonic series	Urgency ranking
146.8	Random	High
523.3	10% irregular	Moderate
523.3	Regular	Low

Table 8.2 **Warning-to-system-failure mapping and proportion of trials in which tracking control is lost**

Subject	Warning		
	High	Moderate	Low
1	0.90	0.50	0.10
2	0.90	0.10	0.50
3	0.50	0.90	0.10
4	0.10	0.90	0.50
5	0.10	0.50	0.90
6	0.50	0.10	0.90

the tracking task by physically controlling the position of a target with a joystick and responding with a button-press when joystick control of the target was lost. The automated tracking condition required monitoring of the computer performing the tracking task and a button response when control was lost. The loss of tracking control (probability of failure) determined the situational urgency level.

The following hypotheses were tested in this experiment.

1 Subjective assessments will reveal that subjects are able to rank the perceived urgency of the warnings.
2 Faster reaction times will occur during the manual tracking condition. Faster reaction times will also occur in response to the most urgent warning (perceived).
3 EEG alpha power (8–12Hz) and theta power (4–7Hz) will be increased during the automated tracking condition. An interaction is expected to occur between tracking condition and urgency of the auditory warning, with the largest alpha and theta power increases occurring during the automated tracking condition in response to the least urgent warning (perceived).
4 An increase in attentional engagement will be reflected in an EEG frequency, and in ERPs in response to the most urgent warning (perceived).

Method

Subjects

Subjects consisted of six volunteer students who ranged in age from 20 to 30 years, with an equal number of males and females. All subjects were right-handed and reported normal or corrected-to-normal vision. All subjects reported normal auditory functioning and no history of neurological problems which could interfere with the recording of EEG. Subjects were treated in accordance with the Ethical Principles of Psychologists and Code of Conduct (APA, 1992).

Apparatus and Stimuli

EEG and ERPs were collected with an Electro-Cap International elastic head cap consisting of 22 recessed electrodes arranged according to the standardised International 10-20 placement system (Jasper, 1958). These electrophysiological signals were recorded on to an optical disk of a Cadwell topographical brain mapping system (Model No. Spectrum-32), a standard EEG recording device which can record continuous EEG and is equipped with software capable of extracting and analysing ERPs from within the EEG record. The tracking task was based on the Multi-Attribute Task Battery (Comstock and Arnegard, 1992) and was displayed on a Magnavox 35.5cm diagonal colour computer monitor (Model No. 9CM062 074I). The auditory warning signals were generated with an Ensoniq Digital Auditory Wave Synthesizer (Model No. ESQ-1) controlled through MIDI via an IBM PC/AT microcomputer. The experiment was conducted in a sound-attenuated room with a reverberation time of approximately 120ms. Ambient noise level was 50dBA (broadband). Signals were presented through one loudspeaker (Acoustic Research Model No. PP570) which was positioned directly in front of the subject at a typical distance of 1m. All sound pressure levels were determined by a Simpson sound level meter (Model No. 866). Amplitude envelope and frequency parameters were measured with a Philips analogue storage oscilloscope (Model No. PM-3266) and a Nicolet Scientific Corporation Mini-Ubiquitous FFT-Computing Spectrum Analyzer (Model No. 446AR).

As shown in Table 8.1, the low urgency warning was a 523.3Hz pulse waveform with all harmonics (1–28) present at approximately equal powers. A pulse waveform on the Ensoniq ESQ-1 Synthesizer was essentially a square wave with a very low duty cycle (that is, a series of impulses). The moderate urgency warning comprised a 523.3Hz pulse waveform and a 1141.7Hz pulse waveform sounded simultaneously to approximate a 10 per cent upward shift of the even harmonics of the 523.3Hz waveform. The high urgency warning comprised a 146.8Hz pulse waveform, a 312.5Hz pulse waveform and a 415.3Hz pulse waveform sounded simultaneously. The 312.5Hz waveform simulated a 10 per cent upward shift in the even harmonics of the 146.8Hz waveform, and the 415.3Hz waveform simulated a 50 per cent upward shift in the even harmonics of the 146.8Hz waveform. All the harmonics from all the components were present in equal powers. Each stimulus had a standard amplitude envelope, was presented at approximately 75dB(A) (that is, 25dB(A) above the ambient noise level of 50dB(A)) and consisted of a 1.2s burst of six pulses. Each pulse lasted 200ms and had a linear rise and fall time of 20ms. The first pulse was always attenuated by 3dB(A) relative to the other pulses in the burst, in order to reduce startle reactions.

Design and Procedure

A repeated measures 3 x 2 (urgency of the auditory warning x tracking condition: (manual or automated tracking)) design was employed. However the pairings of auditory warning and the probability of tracking system failure (0.90, 0.50 or 0.10) were systematically counterbalanced between subjects so that subsequent analyses could be performed using both the perceived and situational urgency scales.

Prior to the experimental session, an elastic electrode head cap was fitted onto the subject's head in order to permit the recording of EEG. The electrode sensors in the cap were placed over the prefrontal (Fpz, Fp1, Fp2), frontal (Fz, F3, F4, F7, F8), central (Cz, C3, C4), parietal (Pz, P3, P4), temporal (T3, T4, T5, T6) and occipital (Oz, O1, O2) regions of the scalp. Additional electrodes were placed on the mastoid prominences as an electrical reference for the other sensors. Following sensor placement, the subject was seated in an experimental chamber, and baseline EEG measures were recorded for five minutes. The subject was then asked for his or her subjective assessments of the warnings. In the first procedure, the subject was instructed to identify the most urgent and least urgent warning. This procedure yielded a rank ordering of the warnings. In the second procedure, the subject was instructed to provide subjective urgency ratings by estimating the urgency of each warning on a rating scale having zero labelled 'not urgent' and 100 labelled 'very urgent'. After assessing the warning urgency levels, the subject was given the assigned warning-to-system-failure mapping and was instructed to participate in a tracking and reaction time task while EEG was recorded.

The tracking task consisted of manual and automated conditions and required that a circular target be kept within a rectangular boundary. During each tracking condition, each auditory warning was assigned to a probability (0.90, 0.50 or 0.10) of the tracking system failure. For example, in a case of direct urgency mapping, the high urgency (perceived) warning indicated a 0.90 probability of system failure, the moderate urgency warning indicated a 0.50 probability of system failure and the low urgency warning indicated a 0.10 probability of system failure. In reverse mapping, for example, the high urgency warning indicated a 0.10 probability of system failure. Direct mapping, reversed mapping and all other possible combinations of mapping between perceived urgency and situational urgency were incorporated into the experimental design. Each subject was presented with only one set of warning-to-system-failure probabilities, as shown in Table 8.2.

Each condition of the tracking task lasted 15 minutes and a three-minute practice period was given to the subject before each of the actual conditions, which were presented in counterbalanced order. During the manual condition, subjects were required to track manually a circular target on a video screen with a joystick and to perform a reaction time task. During the automated condition, subjects were required to monitor computer tracking of the circular target and to perform only the reaction time task.

Throughout the tracking task, an auditory warning signifying the probability of system failure was presented at 30s intervals, and the circular target changed in shape to a square. If the square target then became uncoupled from joystick control and started to drift outside the rectangular boundary, the tracking system had failed. Subjects were required to press the button on the joystick in order to resume tracking. Reaction time was recorded from the time the tracking system failed until the push-button response. An auditory ERP was recorded in response to each warning to assess the immediate physiological response to each warning. Continuous EEG was recorded during both the automated and manual tracking conditions to assess the current physiological responses of attention to the three warnings.

After the completion of the tracking task, the auditory warnings were presented again, and the subject was instructed to ignore the meanings which were assigned to the warnings during the tracking task. As before, the subject was first asked to identify the most urgent and least urgent warning and was then asked to provide subjective urgency ratings by estimating the urgency of each warning on a rating scale having zero labelled 'not urgent' and 100 labelled 'very urgent'.

Results

Subjective Assessments

Subjects were presented with the three auditory warnings at the beginning and end of the tracking experiment and were asked to provide subjective urgency ratings of each warning as well as to identify the most and least urgent warning.

Ratings of urgency level   Figure 8.1 displays the mean urgency ratings and 95 per cent confidence intervals for each warning. These means were computed from the subjective assessments and were used to determine the extent to which subjects were able to distinguish among the perceived urgency levels of the auditory warnings. A significant difference existed among subject ratings of the perceived urgency before the experimental session (F (2,10) = 6.95; p < 0.05). These results are consistent with those found in the Edworthy *et al.* (1991) study. A Newman–Keuls post hoc analysis revealed a significant difference between the high and moderate urgency warnings and between the high and low urgency warnings during the ratings made before the experimental session. No significant results were found among the warnings for the ratings made after the experimental session.

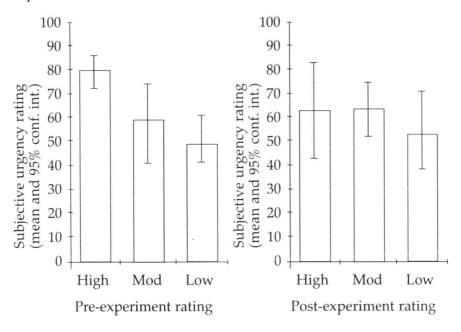

Figure 8.1   Subjective urgency ratings

Ranking of urgency level   An additional procedure used to evaluate subjective perceptions of urgency involved asking subjects to identify the most urgent and least urgent warning. A Friedman test for matched samples was performed on the subjective rankings of the warnings and revealed a significant difference among the rankings before the experimental session (F_r (3,6) = 7.0; $p < 0.05$). Subject rankings were in agreement with the perceived urgency scaling. A non-significant difference was found after the experimental session (F_r (3,6) = 1.333; $p > 0.05$). These data demonstrate that subject rankings of warning urgency level prior to the experiment were consistent with Edworthy's results and our perceived urgency scaling, but that the experimental mappings altered subsequent rankings.

Performance

Reaction times were analysed by comparing the median reaction time to the high, moderate and low urgency warnings in both tracking conditions with a 3 x 2 (urgency x tracking condition) repeated measures ANOVA. Reaction times did not differ significantly in response to perceived or situational urgency level. Reaction times were significantly faster (F (1,5) = 6.63; $p \leq 0.05$) in response to both perceived and situational urgency during the automated tracking condition (461ms) as compared to the manual tracking condition (828ms).

Physiological Signals

A summary of the physiological results based on the perceived and situational urgency manipulations is provided in Table 8.3. All of the physiological indices represented ratio-scaled data with known variability and were analysed using repeated measures ANOVA (SPSS, Inc., Chicago, Ill., USA) which offers additional control of between subject variability. To identify specific differences found using the ANOVA procedure, orthogonal contrasts were employed. The orthogonal contrasts were *a priori* procedures requiring selection of differences of interest before analysis, unlike less conservative post hoc tests. Specific physiological results are outlined in the following paragraphs.

EEG absolute power   Based on the work of Makeig *et al.* (1990) on alertness, EEG data from the vertex region were chosen. Data from sites Cz (mid-line central), Fz (mid-line frontal) and F3 (left frontal) were analysed by computing the absolute power of the theta, alpha and beta frequency bands in response to each warning during both

Table 8.3   Summary of significant physiological results based on two urgency manipulations

Measure	Perceived urgency	Situational urgency
EEG absolute power	Alpha at Fz and F3: difference by tracking condition (manual versus automated)	Alpha at Fz and F3: difference by tracking condition (manual versus automated)
		Alpha at Cz and Fz: difference by urgency level
EEG frequency ratio	Cz: interaction between urgency level and tracking condition	Cz: difference by urgency level
$\dfrac{\text{beta}}{\text{(alpha +theta)}}$		F3: difference by tracking condition (manual versus automated)
Auditory ERP	No significant results	N2 at Cz: interaction between urgency level and tracking condition
		N1 ~ P2 at Cz: difference by urgency level

the manual and automated tracking conditions. These data were compared by way of a 3 x 2 (urgency x tracking condition) repeated measures ANOVA in order to identify any significant brainwave changes occurring in response to the warnings during the manual and automated conditions of the tracking task. Previous research suggests that alpha is the frequency which best reflects attentional differences (Mulholland, 1983). Analyses involving theta and beta revealed non-significant results at those sites.

In the analysis of situational urgency, alpha power at site Cz was computed and revealed a significant difference in response to the warning urgency level (F (2,10) = 6.57; $p < 0.05$). A contrast analysis revealed that increased alpha activity occurred in response to the moderate urgency warning. Alpha at Fz revealed significantly more alpha power in the automated tracking condition than in the manual tracking condition (F (1,5) = 7.26; $p < 0.05$). Alpha power at Fz also revealed a significant difference in response to the warning urgency level (F (2,10) = 5.39; $p < 0.05$). A contrast analysis revealed that the most alpha activity occurred in response to the low urgency warning. Alpha power at F3 revealed significantly more (F (1,5) = 7.59; $p < 0.05$) alpha activity occurring during the automated tracking condition.

In the analysis of perceived urgency, alpha power at site Fz revealed a significant difference between the manual and automated tracking conditions (F (1,5) = 7.26; $p < 0.05$) with more alpha activity occurring during the automated tracking condition. The power of alpha at F3 revealed that there was a significant difference between the manual and automated tracking conditions (F (1,5) = 7.59; $p < 0.05$) with more alpha occurring during the automated tracking condition.

EEG frequency ratio   EEG data from site Cz were analysed further by determining the ratio of fast to slow brainwave activity, beta/(alpha + theta). This ratio has been shown to be an effective index of operator engagement by Pope *et al.* (1994).

In the analysis of situational urgency, a 3 x 2 (urgency x tracking condition) repeated measures ANOVA revealed that significant attentional differences occurred at site Cz in response to the urgency level (F (2,10) = 4.29; $p < 0.05$). A contrast analysis revealed that the lowest index of attentional engagement occurred in response to the moderate urgency warning. Alpha power at site F3 revealed a significant difference between the manual and automated tracking conditions (F (1,5) = 6.51; $p \le 0.05$) with more alpha activity occurring during the automated tracking condition.

In the analysis of perceived urgency, a 3 x 2 (urgency x tracking condition) repeated measures ANOVA revealed a significant interaction effect between the warning urgency level and the condition of the tracking task at site Cz (F (2,10) = 4.60; $p < 0.05$). A contrast analysis revealed that subjects were the most alert in response to the high urgency warning during the automated tracking condition and were the least alert in response to the low urgency warning during the manual tracking condition.

Event-related potentials   ERP data were extracted from the EEG, and the average auditory- evoked potential at site Cz for the high, moderate and low urgency warnings (perceived and situational) in each tracking condition was computed. Early sensory and perceptual components of the ERP occurring before 200ms, as well as a later cognitive component occurring at about 400ms, were analysed. A 3 x 2 (urgency x tracking condition) repeated measures ANOVA computed on the amplitude difference between the early sensory components N1, the first negative peak, and P2, the second positive peak, revealed a significant difference (F (2,10) = 4.18; $p <$ 0.05) between the responses to situational urgency levels. A contrast analysis revealed that the amplitudes which occurred in response to the low and moderate urgency warnings were significantly different from one another, with the largest amplitude differences occurring in response to the moderate urgency warning. A 3 x 2 (urgency x tracking condition) repeated measures ANOVA on the later cognitive component N2, the second negative peak, revealed a significant interaction (F (2,10) = 8.09; $p <$ 0.01) between the level of situational urgency and the condition of the tracking task. A contrast analysis revealed that the largest amplitudes occurred in response to the high urgency warning during the manual tracking condition and in response to the low urgency warning during the automated tracking condition. No significant results were found for responses made to perceived urgency levels.

General Discussion

The first hypothesis under test was that subjects would be able to rank the perceived urgencies of the warnings. In this study, subjects demonstrated the ability to rank the perceived urgency of warning sounds which varied only in their characteristic fundamental frequency and harmonic structure. However these rankings, which were consistent with Edworthy *et al.* (1991) prior to the tracking experiment, were altered following the performance of a tracking task which reassigned urgency in a random fashion, without respect for the inherent acoustical characteristics. Thus it would appear that, in learning the task, subjects reassigned urgency on the basis of task demands and ignored previous urgency rankings based solely on the sound characteristics. Although this adaptability is admirable, the process of suppressing natural reactions to auditory stimuli can become prohibitive in complex environments, such as in aircraft cockpits, hospital intensive care units or in other monitoring and control workplaces.

The second hypothesis addressed reaction times in response to perceived and situational urgency, and to manual versus automated tracking conditions. Slower reaction times in the manual tracking condition were found for both perceived and situational urgency. Although it was hypothesised that faster reaction times would occur during the manual tracking condition, the present results are consistent with prior in-house results which also showed faster reaction times in the automated tracking condition. One explanation for this phenomenon may be that an experimental session which lasts less than half an hour may not provide enough time for subject performance decrements to appear. It is also possible that subjects produce faster reaction times during the automated tracking condition because they are not occupied with any other task (that is, the tracking task may interfere either physically or cognitively with the push-button response). Finally the lack of an urgency effect on reaction time may be due to the fact that the three warnings used in this experiment are rather homogeneous with respect to their range of spectral sound parameters.

The third hypothesis addressed EEG responses to tracking task condition and to perceived urgency. Tracking task condition effects revealed increased alpha in the automated tracking conditions for both perceived and situational urgency. Alpha increases are indicative of relaxed, non-processing states, and are associated with decrements of arousal in response to stimuli. Significant increases in alpha level, independent of tracking condition, were found at sites Fz and Cz in response to either the low or moderate probability of tracking task failure for situational urgency. Alpha differences at these sites were not found for perceived urgency. Alpha changes were expected in response to perceived urgency, thus reflecting in the EEG inherent sound characteristics. However only the situational urgency manipulation produced differences in alpha power, suggesting that the task-related urgency overshadowed the sense of urgency due to sound characteristics.

The fourth hypothesis addressed attentional engagement through an EEG frequency ratio (engagement index) and ERPs in response to the warnings. The EEG frequency ratio provides a measure of subject engagement by computing a ratio of frequency powers. The ERP reveals the immediate sensory response and cognitive processing elicited by auditory stimuli. Significant urgency level effects were found for EEG frequency ratio data and for ERP amplitude differences in the situational urgency manipulation, while only a single interaction effect was found in the perceived urgency manipulation. The findings using the EEG frequency ratio and ERPs are similar to the EEG alpha power results in that the situational manipulation of urgency produced significantly different responses to the warnings.

In operational settings the sense of urgency associated with an auditory warning will always be based on both its inherent auditory properties and its situational context. These factors combine to determine the level of attention afforded to a warning and the motivation to take appropriate action. The present study was similar to real-world experience in that it combined the manipulation of task demands with the manipulation of the auditory properties of the warnings. EEG and ERP measures were sensitive to urgency along both dimensions but primarily to the situational urgency manipulation. This finding suggests that EEG and ERP measures of urgency can be used to discriminate between separate factors which constitute perceived urgency.

It would be interesting to investigate auditory warnings with a larger range of spectral sound parameters, since the lowest urgency estimation reported was 50 on a scale of 0 to 100. This suggests that other auditory warnings could be devised in which the sound parameters may characterise very low ratings of urgency. Accordingly different urgency levels on each end of the spectrum may be reflected in physiological as well as reaction time measures. It should also be noted that the true sensitivity of the physiological measures may have been underestimated in the present study owing to limitations in sample size and therefore limitations in statistical power.

Another variation of this study could examine the added value of direct urgency mapping. Such a study might employ one group of subjects who perform the tracking task while being presented with warnings directly mapped to situational urgency while another group of subjects perform the task while being presented with warnings inversely mapped to situational urgency. This type of experimental design would facilitate the comparison of urgency ratings and task performance of subjects exposed to direct mapping with the ratings and performance of subjects exposed to reverse mapping.

Greater knowledge concerning perceived urgency as well as situational urgency will provide information regarding the ways in which warning sound parameters may be manipulated in order to signify urgency levels that promote optimal states of attention and awareness. The use of physiological measures, in conjunction with behavioural and subjective measures, can lead to the development of more effective auditory warning designs.

Acknowledgments

This research was conducted by the Hazards and Error Management (HEM) research program within the Crew/Vehicle Integration Branch of NASA Langley Research Center in Hampton, Virginia 23681, USA. This chapter has been reproduced from its original publication in *Ergonomics*, **38**(11), 2327–40, by the kind permission of Taylor & Francis.

References

American Psychological Association. (1992) 'Ethical Principles of Psychologists and Code of Conduct', *American Psychologist*, **47**, 1597–1611.

Andreassi, J.L. (1989) *Psychophysiology: Human behavior and physiological response*, (2nd edn), Hillsdale, NJ: Lawrence Erlbaum Associates, Publishers.

Axelsson, R. and Stoby, F. (1991) *Investigation of Speech Systems for Ground Collision Warning in Military Aircraft* (Report No. FFA-TN-1991-05), Stockholm: The Aeronautical Research Institute.

Berger, H. (1929) 'On the electroencephalogram of man', *Archives of Psychiatry and Nervous Diseases*, **87**, 511–70.

Bertone, C.M. (1982) 'Human factors considerations in the development of a voice warning system for helicopters', in C.W. Conner (chair), *Behavioral Objectives in Aviation Automated Systems Symposium, Proceedings No. P-114*, Anaheim, CA: Aerospace Congress & Exposition.

Comstock, J.R. and Arnegard, R.J. (1992) 'Multi-attribute task battery' (Technical Memorandum 104174), Hampton, VA: NASA Langley Research Center.

Doll, T.J. and Folds, D.J. (1985) 'Auditory signals in military aircraft: Ergonomic principles versus practice', in R.S. Jensen and J. Adrion (eds), *Proceedings of the Third Symposium on Aviation Psychology*, Columbus, OH: Ohio State University Department of Aviation.

Doll, T.J., Folds, D.J. and Leiker, L.A. (1984) 'Auditory information systems in military aircraft: Current configurations versus the state of the art' (Report No. USAFSAM-TR-84-15), Atlanta, GA: Georgia Institute of Technology.

Edworthy, J. (1994) 'Auditory warning design', in M. Vallet (ed.), *Noise as a Public Health Problem*, Arcueil, France: INRETS.

Edworthy, J., Loxley, S. and Dennis, I. (1991) 'Improving auditory warning design: relationship between warning sound parameters and perceived urgency', *Human Factors*, **33**(2), 205–231.

Empson, J. (1986) *Human Brainwaves: The psychological significance of the electroencephalogram*, New York: Stockton Press.

Folds, D.J. (1985) 'Development of a comprehensive voice message vocabulary for tactical aircraft', in R.S. Jensen and J. Adrion (eds), *Proceedings of the Third Symposium on Aviation Psychology*, Columbus, OH: Ohio State University Department of Aviation.

Irving, A. (1981) 'An experimental comparison of operator responses to voice and tone system warnings' (Report No. BAE-BT-12051), Bristol: British Aerospace Dynamics Group.

Jasper, H.H. (1958) 'The ten-twenty electrode system of the International Federation', *Electroencephalography and Clinical Neurophysiology*, **10**, 371–5.

Kerr, J.H. (1985) 'Auditory warnings in intensive care units and operating theatres', *Ergonomics International*, **85**, 172–4.

Kerr, J.H. and Hayes, B. (1983) 'An "alarming" situation in the intensive care ward', *Intensive Care Medicine*, **9**, 103–4.

Kramer, A.F. (1991) 'Physiological metrics of mental workload: A review of recent progress', in D. Damos (ed.), *Multiple-task Performance*, Bristol, PA: Taylor & Francis.

Lazarus, H. and Höge, H. (1986) 'Industrial safety: acoustic signals for danger situations in factories,' *Applied Ergonomics*, **17**(1), 41–6.

Lindsley, D.B. (1960) 'Attention, consciousness, sleep and wakefulness', in J. Field, H.W. Magoun and V.E. Hall (eds), *Handbook of Physiology: Vol. III, Neurophysiology*, Washington, DC: American Physiological Society.

Makeig, S., Elliott, F., Inlow, M. and Kobus, D. (1990) 'Lapses in alertness: brain-evoked responses to task-irrelevant auditory probes' (Report No. NHRC-90-39), San Diego, CA: Naval Health Research Center.

Marshall, R. (1987) 'Automatic voice alert devices (AVAD)', in *Recent Advances in Cockpit Aids for Military Operations*, London: The Royal Aeronautical Society.

Meredith, C. and Edworthy, J. (1994) 'Sources of confusion in intensive therapy unit alarms', in N. Stanton (ed.), *Human Factors in Alarm Design*, Bristol, PA: Taylor & Francis.

Mulholland, T. (1983) 'Attention and regulation of EEG alpha-attenuation responses,' *Biofeedback and Self-Regulation*, 8(4), 585–600.

Munns, M. (1971) 'Ways to alarm pilots,' *Aerospace Medicine*, 42, 731–4.

Patterson, R.D. (1982) 'Guidelines for auditory warning systems on civil aircraft', CAA paper 82017, London: Civil Aviation Authority.

Patterson, R.D. (1989) 'Guidelines for the design of auditory warning sounds', *Proceedings of the Institute of Acoustics*, 11(5), 17–24.

Pope, A.T. and Bogart, E. (1992) 'Identification of hazardous awareness states in monitoring environments', *22nd International Conference on Environmental Systems*, Seattle, WA: SAE International.

Pope, A.T., Comstock, J.R., Bartolome, D.S., Bogart, E.H. and Burdette, D.W. (1994) 'Biocybernetic system validates index of operator engagement in automated task', in M. Mouloua and R. Parasuraman (eds), *Proceedings of the First Automation Technology and Human Performance Conference*, Hillsdale, NJ: Lawrence Erlbaum Associates.

Reinecke, M. (1976) *Experimentelle erkenntnisse zur eignung von stimmwarnsystemen in flugzeugen* (Experimental cognition for qualification of voice warning systems in aircraft) (Report No. DGLR PAPER 76-211), Ingolstadt, West Germany: Flugmedizinisches Institut. (Abstract from NASA RECON Document Reproduction Service, Display No. 77A16531.)

Reinecke, M. (1981) 'Voice warning systems: some experimental evidence concerning application', in K.E. Money (ed.), *AGARD Conference Proceedings No. 311, Aural Communication in Aviation*, London: Technical Editing and Reproduction Ltd.

Rood, G.M. (1989) 'Auditory warnings for fixed and rotary wing aircraft', *Proceedings of the Institute of Acoustics*, 11(5), 59–71.

Rood, G.M., Chillery, J.A. and Collister, J.B. (1985) 'Requirements and application of auditory warnings to military helicopters', *Ergonomics International*, 85, 169–72.

Thorning, A.G. and Ablett, R.M. (1985) 'Auditory warning systems on commercial transport aircraft', *Ergonomics International*, 85, 166–8.

Utsuki, N., Nagasawa, Y., Aramaki, S. and Hagihara, H. (1977) 'Auditory evoked potentials elicited by tones and voice as the alarm signal', *The Reports of the Aeromedical Laboratory*, 18(2), 101–111.

Veitengruber, J.E., Boucek, G.P. and Smith, W.D. (1977) 'Aircraft alerting systems criteria study, Vol. I: Collation and analysis of aircraft alerting systems data' (Report No. AD-A042 328), Seattle, WA: Boeing Commercial Airplane.

Wheale, J.L. (1981) 'The speed of response to synthesized voice messages', in K.E. Money (ed.), *AGARD Conference Proceedings No. 311, Aural Communication in Aviation*, London: Technical Editing and Reproduction Ltd.

Wheale, J.L. (1982) 'Performance decrements associated with reaction to voice warning messages', in G.H. Hunt (chair), *AGARD Conference Proceedings No. 329, Advanced Avionics and the Military Aircraft Man/Machine Interface*, London: Technical Editing and Reproduction Ltd.

Wheale, J.L. (1983) 'Evaluation of an experimental central warning system with a synthesized voice component,' *Aviation, Space, and Environmental Medicine*, 54, 517–23.

PART IV
ALARMS IN THE BROADER CONTEXT

9 Investigations of Alarm Mistrust under Conditions of Varying Alarm and Ongoing Task Criticality*

JAMES P. BLISS, *The University of Alabama in Huntsville*

Introduction

The tendency for designers and users of automated systems to rely on emergency signalling systems has increased in recent times, largely because of increased use of automated systems. Unfortunately this dependence may lead to problems. As noted by Breznitz (1983), 'With the rise of sophisticated early warning systems, false alarms are inevitably on the increase, and their psychological impact may well turn out to be the most vulnerable link of many warning systems' (p.xiii). More recently, Edworthy and Adams (1996) stated that 'Just as for visual warnings, the temptation with auditory warnings is to overuse them. Designers tend to operate with a "better-safe-than-sorry" philosophy which dictates that, if there is any doubt, a warning should be implemented' (p.101). Several factors have contributed to such overreliance, including increased societal concerns about consumer safety, and improved sensor technology that enables sensitive signalling systems to be easily implemented.

In complex task environments, operators faced with a large number of sensitive visual and auditory signals have frequently disabled those systems because of false alarms. Sorkin (1988)

*Parts of this chapter relating to ongoing task criticality were adapted with permission from *Proceedings of the Human Factors and Ergonomics Society 39th Annual Meeting* (1995). Copyright 1995 by the Human Factors and Ergonomics Society. All rights reserved.

acknowledged that this problem exists for aviation; in recent years, certain cockpit alarm systems (such as the Traffic Collision Avoidance System and the Ground Proximity Warning System) have gained a reputation for emitting false alarms, leading pilots to mistrust and even disconnect them. The same trend has been noted for alarms in intensive care units. In 1985, McIntyre showed that medical personnel frequently disable alarm systems because of disproportionate numbers of false alarms. Another example was described by Seminara *et al.* (1977) with regard to nuclear power plant control panel alarms:

> In many cases alarm set-points were known by operators to be too sensitive to normal transients. As a consequence slight deviations or transients, thought of as normal, would set the alarm off even though no further operational action was required. Maintenance or calibration operations often caused recurring alarms that were a nuisance. The net result of the many false alarms is a 'cry-wolf' syndrome which leads to a lack of faith in the system and a casual attitude towards the constant presence of certain alarms. (p.245)

One final example of alarm mistrust involves a series of incidents involving aircraft separation by Air Traffic Control in and around Atlanta Hartsfield Airport in the late 1970s (cited in Billings, 1991). Upon investigation, the National Transportation Safety Board reported:

> The flashing visual conflict alert [on the Air Traffic Controller's panel] is not conspicuous when the data tag is also flashing in the handoff status. The low altitude warning and conflict alerts utilize the same audio signal which is audible to all control room personnel rather than being restricted to only those immediately concerned with the aircraft. This results in a 'cry-wolf' syndrome in which controllers are psychologically conditioned to disregard the alarms. (p.100)

The mistrust that stems from multiple false alarms, evident in the above examples, has led researchers to acknowledge the need to investigate human mistrust of alarm systems, or the 'cry-wolf' effect. In particular, scientists have struggled to understand the physiological bases for the cry-wolf effect, as well as the behavioural and system-related factors that influence alarm mistrust (Breznitz, 1983; Bliss, 1993; Getty *et al.* 1995; Kerstholt *et al.* 1996).

Differentiation of the Terms 'Alert', 'Alarm' and 'Warning'

One of the most important elements of any systematic research effort is the establishment of commonly accepted terminology, so that scientists may communicate with each other and with those in applied domains. A consideration of the emergency signal literature shows that such signals are referred to by a variety of names. Scientists and laypeople alike have been inconsistent with regard to the terminology they use when referring to auditory alarms. Some refer to them as 'alarms', whereas others may use the term 'alert' or 'warning'. Furthermore, in specialised task environments such as nuclear power, aviation or medicine, other terms, such as 'telltales', 'flags' or 'advisories' may be adopted, in addition to numerous qualifications of the terms 'alarm', 'alert' or 'warning'. For example, Marshall and Baker (1994) refer to 'first up', 'prime cause' and 'nuisance' alarms. In an attempt to clarify matters, classification schemes have been proposed by some researchers (cf. Baber, 1994, Stanton, 1994). However, these systems do not address general emergency signal terminology per se. For example, Baber's (1994) system of telltales, advisories and warnings is meant to apply specifically to in-car environments. Stanton's (1994) classification system does not address terminology so much as it presents a taxonomy of activities following the onset of an alarm.

In the current chapter, we choose to adopt a classification system for general emergency signal terminology presented earlier by Bliss and Gilson (1998). That system specifies the following.

1 An 'alert' generally indicates that a hazardous occurrence *is expected* soon if nothing is changed.
2 An 'alarm' is a signal indicating that a hazardous event *is occurring* and requires an immediate response.
3 A 'warning' generally signifies that a hazardous incident *may occur* if certain circumstances prevail.

Following from this categorisation, alerts tend to include an element of advanced notice, so that there is some time before action must be taken by the operator. Alarms, by contrast, are more urgent, and mandate an immediate situational fix. Finally, warnings are largely based upon a contingency and generally include written labels, about which much has been written and studied (cf, Wogalter *et al.* 1987).

Considering the differences among signal types is not just an intellectual exercise. In practice, the kinds of behaviour that happen in response to alerts, alarms and warnings may logically be expected to differ (Bliss, 1993). For example, humans frequently ignore written warnings, so that compliance is usually the exception rather than the rule (Rodriguez, 1991). In fact this is a great enough problem to warrant application of various risk analysis and utility models for its understanding (see Edworthy and Adams 1996, for a pertinent discussion of such models). The factors influencing such non-compliance are legion, and may include improper warning placement, urgency mapping, font, colour, size or other specifications. An interesting issue is that, because most written warnings prescribe corrective action rather than remedial action, they may be associated in the user's mind with a lack of danger. Therefore the user may not trust the warning to provide critical information.

In contrast to written warnings, the default action for an alarm is usually some sort of behavioural response such as alarm cancellation or remediation of the problem. In fact it has even been suggested that humans are physically predisposed to respond to alarming sounds, such as a scream (Bliss 1993), an idea supported by the vast amount of work concerning the startle effect. Interestingly in many cases the designers of alarms go out of their way to make alarms loud and annoying, so that the operator has no choice but to detect them. Therefore non-compliance is usually not an issue of non-detection (as it often is with written warnings), but may be instead due to cognitive factors such as preoccupation with other tasks or a lack of faith in the alarm system.

Finally the observed action for alerts may vary widely, and will almost certainly be dependent upon the task performance situation. For example, if the consequences of non-response are high (such as when flying in bad weather), an alert from Air Traffic Control concerning making altitude adjustments may be responded to more quickly by the pilot than the same alert issued in fair weather conditions.

The concept of an alert is represented particularly well by the research of Edworthy *et al.* (1995) regarding trendsons. Edworthy *et al.* describe trendsons as precursors to alarms. Trendsons enable operators to benefit from advance notice before a crisis occurs. From a design standpoint, the use of alerts may be preferable to the use of alarms, because the operator has a greater opportunity to respond to or avoid the crisis (Gilson and Phillips, 1996). However alerts or trendsons that occur too far in advance are often perceived as irrelevant or false, in which case operator trust in the signalling system may be compromised (Breznitz, 1983; Bliss, 1993).

Research Concerning Alarm Mistrust

From the above discussion it follows that mistrust may occur for alarms, alerts or warnings, yet only a small group of researchers have concerned themselves with trying to understand the mechanics of emergency signal mistrust. Even in the case of alarms (where a lack of trust is clearly observable because of normal response behaviour that is not demonstrated), few researchers have concerned themselves with the lack of trust that results from frequent false alarms (the 'cry-wolf' effect). Instead most have investigated false alarms from a detection standpoint, using tools such as Signal Detection Theory to investigate human abilities to detect signals (Green and Swets, 1966). Although Signal Detection Theory provides for a way to measure response behaviour (the 'receiver operating characteristic') and may include mistrust in the conceptualisation of beta (ß), it does not enable researchers to examine the cognitive mechanisms of mistrust in detail. Only recently have researchers begun to utilise other methods to examine mistrust (Breznitz, 1983; Paté-Cornell, 1986; Bliss, 1993).

As noted in other work (Bliss *et al.*, 1996), the decision process leading to an alarm response decision is highly dependent on operator trust. If an operator mistrusts the alarm system because of frequent false alarms, responding may decrease, as some authors allow (Breznitz, 1983), or it may cease altogether, as others state (Paté-Cornell 1986). Breznitz refers to alarm mistrust in terms of a loss of credibility that the alarm system suffers. He states, 'The credibility loss following a false alarm episode has serious ramifications to behavior in a variety of response channels. Thus, future similar alerts may receive less attention' (p.11).

The tendency for an unreliable alarm system to mislead and confuse operators is compounded when true and false alarms are indistinguishable, as they often are in complex task environments. This is so because of the value of additional information upon which an operator may rely for information concerning the underlying cause of an alarm. Breznitz (1983) describes this additional information as 'extrinsic'; others speak of 'redundancy' within an automated alarm system (Selcon *et al.*, 1995). Bliss *et al.*, (1996) showed the value of such additional information clearly. In their experiment, responses to marginally reliable alarms were made more quickly and appropriately when participants could refer to an ancillary display showing information about the reliability of the alarm system or validity of each individual alarm. Presumably the presence of an additional display helped bolster the participants' trust in the alarm system.

Several researchers have discussed trust within the context of the human–automation relationship. In 1993, Barber outlined trust in three dimensions: fundamental laws, technical competence and fiduciary responsibility. Together these seem to be related to the types of expectations involved in interpersonal relationships. He indicated that persistence of these dimensions would all be expected in a trust relationship. In describing interpersonal relationships, Rempel *et al.*, (1985) listed trust expectations as predictability, dependability and faith. Based on this framework, Muir (1994) proposed a theory outlining the role of trust in the human–machine relationship. In her theory, trust depends upon four dimensions: (1) an assumption of natural and social order; (2) the expectation of performance consistency, stability and desirability; (3) an appreciation for underlying qualities of methodology; and (4) an understanding of the initial intentions for the system.

An important consideration in this explanation is the issue of human faith, which takes a fundamental role after extended experience in the relationship (Lee and Moray, 1992). Within the context of alarm reactions, faith may be synonymous with trust. The 'cry-wolf' effect, as witnessed in a human–machine interaction, is a good example of the loss of that faith, following noted alarm system performance inconsistencies.

Although most alarm research has not concerned the lack of trust in human–alarm relationships, the necessity of studying alarm mistrust has been acknowledged by researchers. Tyler *et al.* (1995) recently compiled a report detailing the frequency with which military aviation incidents and accidents were related to the failure of an alarm system to inform operators, as well as alarm mistrust. Their report shows that the 'cry-wolf' effect is a major contributor to accidents involving automated systems. On the basis of this information, Gilson *et al.* (1996) have recommended new training strategies to counter alarm failure and mistrust.

The fact that alarm failure or mistrust can lead to accidents is not new information. Aviation safety databases such as the Aviation Safety Reporting System (ASRS) routinely specify alarm problems as causes of accidents. In 1991, Billings published a report that concerned the growth of automation in civil aviation environments. In that report are details surrounding a number of serious accidents involving civil airliners. Of these, many were due to malfunctioning or misinterpreted alarm systems. The present author is currently compiling a list of recent civilian air transport incidents and accidents attributable to alarm mistrust and failure (Bliss, in preparation).

Along with the growing acknowledgment of the problem, there have been a number of researchers engaged in laboratory research to examine factors related to alarm mistrust. Some of these efforts have focused on understanding the nature of alarm mistrust, while others have concentrated on improving alarm response tendencies. According to Breznitz (1983), the prevalence of false alarms results in a mistrust of subsequent alarms that influences response performance. He notes that the widespread use of highly sensitive warning systems has made alarm mistrust (the 'cry-wolf' effect) a prevalent phenomenon. A group of studies conducted by Breznitz (Breznitz, 1983) concentrated on many aspects of the 'cry-wolf' effect, including the role of operator expectancy (readiness), the effect in divided attention (dual-task) situations and certain behavioural changes that precede and follow the effect. Breznitz concludes his work with a set of 25 propositions for minimizing the false alarm effect, based on the research he and his colleagues conducted. These propositions are meant to offer designers and trainers a sort of guide for minimising the 'cry-wolf' effect. Breznitz's work is important, because it considers theoretical and psychophysiological implications of the 'cry-wolf' effect; however he does not attempt to utilise task response measures such as alarm response frequency, speed, accuracy or appropriateness.

Paté-Cornell (1986) later attempted to use statistical models to predict response behaviour to unreliable alarms. An important element of her model is the predictability of response patterns following a sequence of false alarms. Her work is perhaps the first to attempt to predict operator response patterns mathematically, taking into account alarm mistrust. Her model remains to be validated by any performance data, however. Both Breznitz (1983) and Paté-Cornell (1986) have been successful in showing that the 'cry-wolf' effect exists in the form of increased physiological stress and predicted response behaviour, yet a critical element is still missing: does alarm mistrust manifest itself in degraded alarm or ongoing task performance? Recently scientists have begun to formulate an answer to that question.

In the last few years, researchers such as Getty *et al.* (1995), Bliss *et al.* (1995), Dunn (1995) McDonald *et al.* (1995) and Bliss (1997) have used dual-task paradigms to investigate and improve alarm response performances when a percentage of the alarms are known by the operator to be false. Getty *et al.* (1995) utilised 'receiver operating characteristics' to evaluate undergraduate students' responses to alarm systems with varying positive predictive values (PPVs) or false alarm rates. They found that students responded more slowly to alarm systems with a lower PPV, presumably owing to mistrust. This replicates earlier findings by Bliss (1993), who noted similar indices of mistrust in the form of decreased alarm response accuracy, appropriateness and frequency.

Bliss *et al.* (1995) showed that the tendency to exhibit degraded responding to an unreliable alarm system could be counteracted by presenting participants with updated information about an alarm system, or by increasing the physical salience of the alarms. These authors found that the presence of hearsay information led college students to adopt greater faith in a marginally reliable alarm system. As a result, they responded more frequently to alarms generated by that system. Dunn (1995) investigated the influence of primary and secondary task workload on responses to a marginally reliable alarm system, noting that performances on the alarm and ongoing task may be predictable by using Wickens's (1984) multiple resource theory to determine when attentional shifts will or will not occur. Dunn's results supported her hypotheses: high primary and secondary task workload levels compounded the 'cry-wolf' effect, resulting in further decrements in response parameters.

McDonald *et al.* (1995) investigated the role that multiple alarms play to modify operator confidence in the alarm system. They presented participants with alarms, 50 per cent of which were true alarms. At the same time, they presented zero to five collateral alarms. McDonald *et al.* found that, when inexperienced operators judge the validity of an alarm, they may be influenced by the presence or absence of alarms that activate at the same time as the original alarm. This suggests that designers may improve perceptions of alarm system reliability by the judicious use of redundant alarm stimuli. Bliss *et al.* (1996) investigated the utility of adding a redundant source of alarm validity and reliability information. They found that having an additional source of such information was indeed helpful to participants. Those with alarm system reliability information responded more frequently than those without, and those with additional alarm validity information responded less frequently but more appropriately.

In addition to the above research using college students as participants, Bliss (1997) conducted research with pilots which showed that the opportunity to react to alarms using vocal commands increased response rates to a marginally reliable alarm system and improved performance on an ongoing complex psychomotor task (the Multi-Attribute Task battery – Comstock and Arnegard, 1992). When given the choice to react to alarms manually (by keypress) or vocally, pilots indicated their preference for the vocal reaction mode; however alarm response times were faster for manual responses.

Most recently work by Kerstholt *et al.*, (1996) has involved investigating skilled operator performance on a simulated ship-handling task, while participants were confronted with an alarm system of marginal reliability. Their findings echo those of Getty *et al.* (1995) and Bliss (1993), showing that responses were less timely and frequent when alarms were unreliable.

The majority of the above investigations required participants to react to alarms while performing a task. The use of a dual-task paradigm to assess alarm response performance represents a promising approach to the false alarm problem. Not only is it ecologically valid, but it offers power and flexibility for operator performance measurement. Research on divided attention suggests that, when people are required to perform more than one task, performance on at least one of the tasks tends to decline, owing to mental resource constraints (Sanders and McCormick, 1993). In complex task situations, alarms constitute an intermittent task which may or may not be acknowledged; in such situations, certain factors related to the ongoing or the alarm task are bound to affect whether or not the operator chooses to acknowledge the alarms.

As the above discussion shows, researchers have addressed several areas affecting reactions to marginally reliable alarm systems, yet one issue that has been largely overlooked when considering operator alarm response patterns is the importance, or criticality, of the tasks being performed. As noted earlier, many researchers are in agreement that responding to an alarm occurs in the context of a dual- (or multi-) task paradigm. Within that context, operators are engaged in a task, and alarms occur as an intermittent secondary task. Breznitz implies that this constitutes a personality variable called the 'dilemma of task orientation'. When this happens, the operator is forced to make a decision to switch attention from the ongoing task to the secondary task, or to ignore the secondary task while maintaining ongoing task performance. As research has shown (Getty *et al.*, 1995; Bliss, 1993), reliability of an alarm system may affect this shifting. Furthermore there are other variables that may also affect it. In particular, the importance of each task may play a part in determining the frequency and timeliness of alarm responses.

Discussion of Task Areas with Varying Alarm and Ongoing Task Criticality

There are many applied situations where alarm and/or ongoing task criticality may be seen to fluctuate. In those situations, multiple-resource theory (Wickens, 1984) suggests that operators may be more or less predisposed to respond to alarms, particularly those that are not completely reliable.

According to theories of selective attention, the decision about how often and when to attend to the tasks available is made by determining the behaviour that will result in the greatest benefit, or least risk (Wickens, 1992). Following from this general logic, Moray (1981) summarised literature concerning laboratory situations where participants are presented with two or more channels, along which events happen at semi-predictable rates. This paradigm bears a resemblance to the dual-task paradigm for alarm reactions discussed earlier, where an 'event' would be a true alarm. Moray concludes that participants form a mental model that guides their sampling (response) behaviour. This mental model is flexible, so that participants will sample channels more often along which events occur more frequently. Participants are especially good at such sampling when they are given an idea about how often to expect events. However Moray (1981) indicates that human memory is fallible, leading to a sort of regression towards the mean, where sampling rates for extreme (high or low) event rates along channels will be more moderate than expected.

It is logical that the criticality (both physical and functional) of the primary task should influence the sampling behaviour of complex task operators. The biggest challenge faced when conducting empirical investigations of criticality is to model that construct acceptably in the laboratory. Upon reviewing the literature, it seems that the notion of criticality may be most closely related to operator perceptions of task utility.

Utility in the context of choice tasks is discussed by Wickens (1992). He notes that, when operators must make decisions under conditions of uncertainty (such as when an alarm system is generating alarms that may or may not be true), they will evaluate each course of action and allocate their effort to the task with the highest payoff or 'expected value'. For the current effort, we choose the term 'criticality' over 'utility' because Edwards (1987) indicates that 'utility' is a subjective phenomenon for each operator. In the following experiments, we manipulate the objective value, or criticality of the ongoing and alarm tasks to determine operators' ensuing performances on each. This effort is important because it is the first of its kind (that we know of) to investigate the criticality issue in conjunction with alarm reliability, and because it will help determine the relationship between utility and value (criticality) in a laboratory alarm system context (see Tversky and Kahneman, 1981).

Aside from the theoretical implications of the present attempt, it is important to recognise that there are a large number of complex task environments where task criticality may fluctuate. Below is a list (by no means exhaustive) of possibilities. Researchers have documented the nature of operator–alarm interactions in each case, and in some cases have gone so far as to investigate variables that moderate the relationship between human operator and alarm system. Included are the following:

- aviation (Patterson, 1982);
- cars (Baber, 1994);
- air traffic control (Hopkin, 1995);
- nuclear power (Seminara et al., 1977; Marshall and Baker, 1994);
- medicine (Kerr, 1985; Meredith and Edworthy, 1994);
- mining (Mallett et al., 1992);
- industrial plants (Duchon and Laage, 1986; Stanton and Baber, 1995); and
- military vehicles (Edworthy and Adams, 1996).

Objectives of the Investigation

In the current research, we presented participants with a dual task situation similar to that used by Sorkin *et al.* (1988), first to determine whether perceived criticality of the alarm task influenced alarm and ongoing task response patterns. We hypothesised that, as alarm reliability decreased, alarm response in terms of speed and accuracy would worsen. We also hypothesised that, as alarm task criticality increased, alarm responses would be more accurate and timely. A third hypothesis concerned ongoing task performance. We predicted that primary task performance would improve as alarm reliability dropped. Finally we expected that, as alarm urgency increased, alarm response time and errors would decrease, regardless of system reliability. These hypotheses were based on the limited resources model of cognitive functioning proposed by Wickens (1992).

Experiment 1: Alarm Task Criticality

When considering criticality of the alarm task, there are two ways that an alarm stimulus may be considered 'critical'. On one hand, the alarm stimulus (auditory, visual, or some other sense) may be physically compelling (hereafter 'salient'). Many researchers have tried to determine which patterns of stimuli are most salient, so that alarms may be constructed to exploit this information (Patterson, 1982; Edworthy, 1994).

Another possibility is that the situation that the alarm is signalling is itself urgent (hereafter 'critical'). In other words, the consequences of not reacting properly to the alarm signal are more significant than those associated with not reacting properly to the same signal in another situation. An example of this type of manipulation might occur in aviation, where an engine fire alarm may occur within easy access of a landing strip in fair weather (low criticality), or the same alarm may occur in bad weather when there is not an easily accessible landing strip within reach (high criticality).

There have been a number of researchers who have examined aspects of salience and criticality. For example, research by Edworthy *et al.* (1991) and Hellier *et al.* (1993) has addressed the problem of multiple alarm response prioritisation, by utilising an urgency mapping technique to exploit alarm salience appropriately. It is hoped that, by utilising such methods, response performances may be optimised without increasing workload. Edworthy and her colleagues have utilised Stevens's Power Law (1957) to quantify the relation between acoustic parameters (including fundamental frequency, spectral composition and harmonic content) and perceived urgency.

The work by Edworthy and her colleagues may provide a way for alarm designers to create alarm systems that facilitate proper response patterns. Specifically those alarms that are more critical or time-sensitive can be coded to elicit a greater feeling of urgency, whereas others that are less pressing may be coded as less urgent. Such coding may help operators adopt the proper response prioritisation sequence and generally elicit faster operator responses (Edworthy and Adams, 1996). Also, by ensuring that the physical salience of the alarm stimulus matches the urgency of the signalled situation (see Haas and Edworthy, 1996), designers may be able to create effective, appropriate alarm stimuli.

In most situations, an alarm's salience is largely independent of its validity (the reliability of the alarm system). However it is likely (especially given the tenets of Wickens's multiple resource theory) that salience and validity may act in an interactive fashion to influence reaction patterns. For that reason, we considered it important to investigate the relationship between these two variables, and to determine how they might interact to influence operator reaction patterns.

Method   Dual-task methodology has been advocated by several researchers as an optimal method for studying intermittent response performance (cf. Damos, 1985; Mastroianni and Schopper, 1986; Putz and Rothe, 1974; Sorkin *et al.*, 1988). When researchers use the dual-task method for examining alarms, alarm responding usually is thought to resemble a selective attention paradigm (Wickens, 1992), where attention in toto is transferred from the ongoing task to the alarm, and back again. Therefore the current experiments used a dual-task paradigm, where participants were required to perform a continuously demanding psychomotor ongoing task. This task was meant to simulate complex primary tasks such as nuclear power control centre operation or aviation. While participants were engaged in this task, they were required to react to alarms that were presented intermittently. As in the real world, such reaction could include responding, or not responding.

It is admittedly difficult to approximate the aesthetics of complex tasks in a controlled setting. Of particular concern to us was the motivation of the participants to perform the experimental tasks. To keep participants motivated, we used monetary incentives combined with extra credit (more about this below).

Participants   The participants in the first experiment included 138 students from undergraduate courses at the University of Central Florida in Orlando. This number was selected on the basis of a power analysis that we conducted using effect size and population variance estimates from prior alarm research (ongoing task estimates were taken from Bliss, 1989; alarm task estimates were drawn from Cardosi and Boole, 1991). The average participant age was approximately 24 years, and each person was screened by demographic questionnaire to ensure adequate vision and hearing. One participant was not allowed to participate because of subnormal hearing.

Participants received extra credit for participation. In addition, a monetary bonus was available, awarded on the basis of the participants' ongoing and alarm task performances (see Table 9.1 for schedule). Participants gained or lost money for each appropriate or inappropriate alarm response, respectively; however the final bonus was awarded only if participants kept their ongoing task speed and accuracy within one standard deviation of baseline (familiarisation) performance. All participants received the monetary bonus.

Tasks It was necessary for the ongoing task to be challenging, so that the effort expended by the participants to perform it (and respond to the alarms) would approximate the effort experienced in many complex task situations. The ongoing task was the Manikin task drawn from the DELTA battery (see Turnage and Kennedy, 1992). A 315-second version of the Manikin task was included in each of two blocks, with each block lasting approximately 10 minutes. An

Table 9.1   Monetary contingencies for alarm responses

True alarm responses

Response type	Note	Warning!	Danger!!
Timely, accurate	+.10	+.15	+.20
Timely, inaccurate	+.05	+.75	+.15
Late, accurate	+.05	+.075	+.15
Late, inaccurate	+.00	+.00	+.00
No response	−.10	−.15	−.20

False alarm responses

Response type	Note	Warning!	Danger!!
Response	−.10	−.10	−.10
No response	+.10	+.10	+.10

IBM-AT compatible 386/25MHz computer generated the Manikin task patterns on a VGA display. Manikin taps spatial ability and mental rotation, and has been shown to predict flight task performance (Turnage *et al.*, 1988). In addition, performance on Manikin stabilises quickly, making Manikin a good choice for repeated-measures research (Jones, 1979). Manikin involves the following (Benson and Gedye, 1963):

- a simulated human figure is presented facing forward or backward;
- participants must determine which hand of the figure (right or left) holds the pattern that matches the pattern at the feet of the figure, and press the appropriate arrow key;
- figure orientation, hand and pattern type are randomly determined for each trial;
- performance is based on the number of correctly matched pairs and latency scores reported in milliseconds from the time the stimulus appears until a response is made.

One advantage of Manikin was that it was continuously demanding. Therefore all alarms occurred while participants were engaged in the ongoing task. Also keeping the ongoing task duration constant allowed the number, duration and temporal separation of the alarms to remain constant across participants.

For the alarm task, stimuli reflecting three levels of urgency were generated, following recommendations by Robinson (1991) and the American National Standards Institute (1991) regarding auditory and visual stimuli characteristics. Alarms were created using Supercard 1.6 software on a Macintosh IIsi 5/80 computer. Concurrent auditory and visual alarms were presented to the participants for redundancy and to ensure detection (Robinson, 1991).

- Visual alarms consisted of a 13.97cm x 22.86cm, rounded-corner, rectangular panel on a white 19.05cm x 24.77cm rectangular background.
- A textual message printed on the alarm panel in black capital letters (48-point Helvetica font) indicated the urgency of the alarm (DANGER!!, WARNING! or NOTE).
- The alarm panel was red, yellow or green, corresponding to a high, medium or low level of urgency (Berson *et al.*, 1981).

The auditory alarms that accompanied visual alarm presentation followed the parameters recommended by Berson *et al.* (1981) and reflected differing urgencies of alarms currently used in modern commercial aircraft.

- The low-, medium- and high-urgency auditory alarms were the Boeing-757 SELCAL chime, the caution siren and the fire bell, respectively.
- Each signal was activated at 65-70 dB(A) once for two seconds, then remained silent.
- The response task was to use the mouse to move the cursor on to a 0.64cm x 0.64cm response panel (labelled 'R') and click the mouse once.
- There were no visual or auditory differences between true and false alarms.
- The distribution of true and false alarms for each group is indicated in Table 9.2.
- Dependent measures included alarm response frequency, accuracy, time and appropriateness.

Participants were presented with detailed instructions regarding the structure and urgencies of the alarms and were given an interactive demonstration of each alarm prior to the beginning of the experiment. No participants indicated confusion regarding the urgencies of the alarms used. As noted above, alarm response measures consisted of frequency, time, accuracy and appropriateness. Frequency refers to the percentage of alarms to which participants responded in each experimental session. Response time was the number of seconds taken by a participant to respond to the alarms following alarm onset. If participants did not respond within 15 seconds, the lack of response was noted and that trial was not included in response speed or accuracy calculation. Responses were accurate if the mouse click occurred on the response button, inaccurate otherwise. Appropriateness refers to the correctness of the response decision (responding to true alarms and not responding to false alarms were both appropriate responses).

When an alarm was activated, the mouse cursor (the arrow) was automatically centred on the screen. Participants were given feedback (an auditory message of 'correct' or 'incorrect') about the appropriateness of their response. If a participant responded to a true alarm or did not respond to a false alarm, the 'correct' message sounded; if a participant responded to a false alarm or failed to respond to a true alarm, the 'incorrect' message sounded. Because there were no differences between true and false alarms, participants were completely dependent upon the auditory feedback for indication of alarm validity and appropriateness of their responses.

Procedure After completing an informed consent form, each participant received familiarisation on the ongoing task and the alarm system. After familiarisation, participants were randomly

Table 9.2 Distribution of true and false alarms within each session according to group

Group	True alarms	False alarms
25% reliability	3	9
50% reliability	6	6
75% reliability	9	3

assigned to one of three alarm reliability groups: 25 per cent, 50 per cent or 75 per cent true alarms. Each group of participants was informed of the actual alarm system reliability in the experimental instructions and this figure was later reinforced in the experimental sessions. Participants then began the first of two experimental sessions. While the participant was engaged in the primary task, alarms were generated on a Macintosh IIsi 5/80 physically stationed 90 degrees to the right of the ongoing task screen. To respond to the alarms, participants had to turn to look at the Macintosh screen, position the mouse cursor on the response panel and click the mouse button.

All participants in each of the three groups participated in two 10-minute task sessions separated by a five-minute rest period. Within each session, participants were presented at various times with four alarms of each urgency, for a total of 12 alarms within each block (24 alarms total per participant across blocks). The presentation time of the alarms was pseudo-randomly generated within blocks on a variable interval (VI) schedule, with the inter-alarm interval ranging from 25 seconds to 75 seconds and the average set at 50 seconds. The sequence of the 12 alarms within each block was random.

Following the onset of each alarm, participants were given 15 seconds to respond, after which the alarm screen disappeared, replaced by a blank, white screen. Depending upon the participant action (response or not) and the nature of the alarm (true or false), one of two spoken messages was presented ('correct' or 'incorrect'). These messages informed participants of the correctness of their decision and indirectly of the alarm validity (whether it was true or false). After the task sessions, the participants were fully debriefed and were given the bonus payment.

Alarm urgency was varied within groups. In real-world complex task situations, true alarms reflect varying degrees of consequences if not responded to in a timely manner. To model such contingencies effectively in the experimental context, each participant was given a 'score' of $5.00 at the beginning of the experiment. During the sessions, money was subtracted from or added to the score, depending upon the appropriateness, timeliness and accuracy of the reaction and the urgency of the alarm (see Table 9.1). Each participant's running total was presented on the alarm screen at all times.

Measures of alarm response performance included the following (for alarm responses, if participants did not respond within 15 seconds, the data point was declared missing):

- the numbers of alarms to which each participant responded, according to session, experimental group and alarm urgency category;
- time taken by a participant to respond to an alarm, measured in seconds from alarm onset to completion of the mouse click;
- whether or not participants positioned the cursor accurately on the respond button;
- the appropriateness of each alarm reaction (making responses to true alarms and withholding responses to false alarms were both considered appropriate reactions).

Discussion of results   We decided to analyse data from the second session only, because most participants indicated that during the first block of the experiment they were still learning the primary and alarm tasks. The results for the first experiment are summarised in Table 9.3. Because establishment of the 'cry-wolf' effect was an important goal of the first experiment, it is important to note that response rates did differ significantly across reliability groups. That shows support for Breznitz's conceptualisation of the 'cry-wolf' effect as a degradation of responding. However, because some participants chose to respond to no alarms, the findings also support Paté-Cornell (1986), who visualised the 'cry-wolf' effect as a total cessation of response. In any event, the presence of an effect for response frequency and appropriateness validates our decision to use the current experimental paradigm for studying alarm mistrust.

Of note also is the fact that response speed and accuracy (for both the ongoing and the alarm task) were insensitive to alarm criticality or reliability, contradicting our hypothesis. This may have been due in part to the experimental instructions and reward structure, which stressed the importance of maintaining response speed and accuracy to all groups. Additionally this finding suggests that participants may have chosen a specific response strategy prior to the start of each experimental session.

Finally, and perhaps most importantly for this experiment, is the fact that participants responded to critical alarms more frequently

Table 9.3   Tabulated results for Experiment 1 by dependent measure

---

Alarm response frequency

Response rates were significantly different across reliability groups; participants in the 25%, 50% and 75% reliability groups responded to 30%, 60% and 82% of the alarms, respectively.

Generally, responders tended to overmatch, responding at a rate higher than the stated true alarm rate.

Approximately 10% of responders responded to all alarms or no alarms.

Regardless of experimental group, the high-urgency alarms were responded to more often than low-urgency alarms.

Alarm response speed

Alarm response speed was not sensitive to differences in alarm reliability.

There was no effect of alarm criticality on response speed.

Alarm response accuracy

Alarm response accuracy was not sensitive to differences in alarm reliability.

There was no effect of alarm criticality on response accuracy.

Alarm response appropriateness

Urgency and reliability influenced response appropriateness in an interactive fashion.

As alarm urgency increased, the appropriateness of alarm responses rose.

Participants in the high reliability group made more appropriate responses than those in the 25% or 50% reliability groups.

Ongoing task

The low reliability group performed the Manikin task faster than the medium and high groups (no difference between medium and high, and no effect of reliability on Manikin accuracy).

There were no effects of alarm criticality on either Manikin speed or accuracy.

---

and appropriately. This indicates that our stimuli were effective; furthermore the fact that response appropriateness showed an interactive effect implies that the criticality of an alarm may moderate an operator's conceptualisation of its reliability, as well as the ensuing responses.

Experiment 2: Ongoing Task Criticality

In most complex task situations, operators are required to perform an ongoing task that may assume varying levels of criticality, and a secondary alarm responding task where the alarms presented may also be more or less critical, depending upon the circumstances. From the first experiment it is evident that alarm criticality may influence the propensity for participants to respond to auditory alarms that are of a less than reliable nature. It is logical to expect factors particular to the ongoing task to influence alarm response patterns also. One particular factor that is highly relevant for alarm situations is ongoing task criticality. Complex task operators are frequently faced with ongoing tasks of varying criticality. For example, an alarm may be activated when an airline pilot is trimming the flaps (low ongoing task criticality) or when the pilot is attending to the glide slope upon landing (high ongoing task criticality).

Following Wickens (1992), it is probably most appropriate to define ongoing task criticality in terms of the consequences for poor performance. For situations where the ongoing task is of low criticality (such as watching television), the consequences of not attending to that task are probably not too disastrous. However, if the primary task consisted of bathing an infant, lack of attention to that task would constitute a grievous error with potentially severe consequences.

In most complex task situations, the criticality of the ongoing task usually fluctuates from periods of relative triviality to periods of relative importance. For example, during most flights, there are periods of calm where pilots, having activated the autopilot, may do little more than monitor communications and scan for traffic. During these times, failure to perform the piloting task is probably not as disastrous as it would be during takeoff and landing, when the piloting task is much more complex.

The current experiment was designed to investigate the influence of ongoing task criticality on performance of the ongoing and alarm task. Following the results of Experiment 1, we predicted that, as primary task criticality increased, alarm responses would be less frequent and less timely. To investigate this, we used essentially the same methodology as we did in the first experiment, with particular exceptions noted below.

- To lend more statistical power to our investigation of ongoing task performance, we consolidated the two measures of Manikin performance (response speed and accuracy) into a single pooled measure: response speed for correct responses.
- We used a one-way design with ongoing task criticality

manipulated by altering the consequences for poor performance between groups.

- The alarms that were presented to participants were created following the same dimensions as the medium-urgency alarms from the prior experiment.
- Within each of two experimental sessions, participants were presented with 10 alarms, 75 per cent of which were true.
- Ongoing task criticality was manipulated by the use of a point score, as described below.

Participants   A power analysis for a one-way, between-participants design was conducted using effect size and population variance estimates from the first experiment reported above. That analysis indicated that 75 participants would be needed for a power of 0.80 at the $p = 0.05$ level. A total of 78 participants (39 males and 39 females, 26 participants in each experimental group) volunteered from undergraduate psychology classes at the University of Alabama in Huntsville. Each person was screened by questionnaire to ensure adequate vision and hearing, and received credit towards their general psychology classes for participation.

Procedure   After completing informed consent and demographics forms, participants were randomly assigned to the low-, medium- or high-ongoing task criticality group. Participants then completed one 30-second practice session of the Manikin task and received instructions regarding the reliability, onset/offset patterns and response contingencies associated with the alarms.

To model alarm task and ongoing task response contingencies effectively in the experimental context, participants were given a score of 50 points at the beginning of the experiment. Penalties were imposed on each participant's score for failure to respond to true alarms, and for responding to false alarms. Conversely points were added to a participant's score for responding to true alarms or not responding to false alarms. Each participant's running total was presented on the alarm screen at all times. Results from the first experiment had supported the utility of this method; participants indicated that they were highly motivated to respond quickly, accurately, and appropriately.

Following familiarisation, participants began the first of two 8.5-minute blocks of the experimental task, accompanied by the alarms. The only aid that participants had when responding was the expected percentage of true alarms (75 per cent), told to them prior to the first experimental session.

As noted above, ongoing task criticality was manipulated between groups. Participants were told at the beginning of the experiment that the number of sessions they were required to complete depended upon their alarm task score; if they did not score high enough after two sessions, a third session would be required. Then participants were told that if their ongoing task (Manikin) performance during Sessions 1 and 2 was too slow or inaccurate, a number of points would be deducted from their alarm score following the second session (5 for low criticality, 10 for medium criticality and 15 for high criticality). Therefore criticality was determined by the consequences for marginal ongoing task performance. In actuality, all participants performed only two sessions.

Discussion of results On the basis of the results from the first experiment, calculations were again performed on the data from the second session only, because those data were deemed most indicative of stable task performance. Tabulated results from the second experiment are included in Table 9.4. It is interesting to note that, as in Experiment 1, alarm response frequency was the most sensitive

Table 9.4 Tabulated results for Experiment 2 by dependent measure

Alarm response frequency

    The low-, medium- and high-criticality groups responded to 89%, 81% and 79% of the alarms presented, respectively.

    When the ongoing task was more critical, participants responded to significantly fewer alarms.

    Tukey post hoc tests indicated that the low-criticality group responded to significantly more alarms than the medium or high groups ($p < 0.05$). However there was no difference between the medium and high groups.

Alarm response speed

    The medium criticality group took the longest to respond, followed by the low- and high-criticality groups, which were not significantly different.

Alarm response accuracy

    Ongoing task criticality was not significantly related to alarm response accuracy.

Alarm response appropriateness

    Ongoing task criticality was not significantly related to alarm response appropriateness.

Ongoing task

    Average response latency (in milliseconds) for correct Manikin responses did not differ across criticality groups.

performance measure, while alarm response accuracy was not sensitive to differences in ongoing task criticality. The fact that participants responded more frequently when the ongoing task was not critical supported our hypothesis, and illustrates the decision-making tradeoffs mentioned by Wickens (1992) with regard to relative value of responding. Also, as in Experiment 1, participants showed the tendency to match according to probability with response rates approximating the true alarm presentation rate (75 per cent); furthermore, there is clear evidence of overmatching here as well as in the first experiment.

The fact that groups did not differ in terms of their alarm response appropriateness suggests that this is a measure that may be sensitive only to differences in alarm reliability. This is logical, given that appropriateness, as we have defined it, relates to decision making about true and false alarms.

Unlike what happened in the first experiment, alarm response speed was dependent upon group assignment. The pattern of differences suggests that participants made their response decisions more quickly for extreme criticality levels, but may have been unsure about how to react given an ongoing task of intermediate criticality.

Finally the fact that there were no group differences of ongoing task performance was perplexing, and may indicate that participants were exerting maximal effort, regardless of group. This position is reasonable because the Manikin task is largely self-motivating.

General Discussion

The results of this research have direct application to complex task performance where operators attending to an ongoing task are confronted with alarms that may be false. It is encouraging to note that the manipulations we used to vary alarm and ongoing task criticality were successful. In Experiment 1, participants responded at different rates to alarms of different criticalities. In Experiment 2, we chose to vary primary task criticality by utilising a point score which was linked to participation time. We initially had our doubts about whether this would work as well as the monetary incentive used in the first experiment; however the results showed the utility of this method.

A related issue involves the appropriateness of using the dual-task method for studying alarm mistrust. Although many researchers (for example, Sorkin *et al.*, 1988) had used this method to investigate responses to true alarms, studying alarm mistrust this way had not been established to any degree. The current work, together with other recent work by Getty *et al.* (1995), Bliss *et al.* (1995), Dunn (1995), and Kerstholt *et al.* (1996), shows the utility of the dual-task method.

First, in both experiments, the frequency of alarm responses was arguably the most sensitive measure of operator performance. Initially we had thought that response speed or accuracy would be more sensitive because such measures as response time have proved sensitive in other alarm response contexts. Perhaps one of the most significant findings was the tendency for participants to match the true alarm presentation rate with their responses. This becomes more consequential in light of the fact that true and false alarms were indistinguishable prior to the reaction decision. The tendency to match may be explainable using probability matching theory (Herrnstein, 1961), or may be a function of participants' ability to make decisions under conditions of uncertainty by considering the payoffs associated with each course of action (Sorkin and Woods, 1985; Wickens, 1992). The latter explanation may be particularly tenable in light of the fact that some participants chose to respond to all or no alarms as a way to maximise their payoff.

These results show the usefulness of Moray's (1981) conceptualisation of a mental model that guides responses. The fact that response patterns showed 'probability matching' is predictable from Moray's ideas; however the tendency of our participants to overmatch is not predictable, given Moray's comments about human failure to sample extreme event rates accurately. Perhaps unreliable alarms represent an exception to Moray's theory.

The extreme responders reinforced the general idea that participants may have constructed their response strategy prior to the start of each experimental session. This is reflected in Experiment 2, because alarm response times were low for alarms that accompanied the low- and high-criticality primary task, but not for those accompanying the medium-criticality ongoing task. If this idea is transferred to applied settings such as aviation, it could imply that operators are strongly influenced by preformed impressions about alarm reliability. This idea has received empirical support elsewhere (see Bliss *et al.*, 1995).

The findings from Experiment 1 suggest that alarm criticality may interact with reliability, in such a way that highly critical alarms may be responded to regardless of reliability. Such an idea opens the door for designers judiciously to exploit urgency to counteract the 'cry-wolf' effect. In fact Bliss *et al.* (1995) have shown the success of this idea. Alternatively the tendency to overmatch in both experiments may indicate that operators are somewhat resistant to alarm systems that have low reliability.

From Experiment 2 it appears that the importance of the ongoing task may affect the overall response decision; if the ongoing task is of great importance, participants may not respond to an alarm as frequently if it may be false. However, if the ongoing task is a trivial one, more frequent responses may be expected. Taken together, the results from both experiments suggest that participants may have a tendency to adjust responses to the true alarm rate, even when that rate is low.

The experiments reported here are only a first step towards understanding alarm mistrust. Given that the 'cry-wolf' phenomenon does exist and can be quantified empirically with performance measures, it is important to determine other factors that might influence its intensity. Some of these are listed below:

- level of operator alertness or motivation;
- past experience with the ongoing task;
- past experience with the alarm system's reliability;
- risk-taking behaviour of the operator;
- perceived 'costs' associated with response or non-response;
- the operator's knowledge of the broader task or alert consequences (that is, 'situation awareness');
- prioritisation schemes inherent in the task structure;
- alarm onset/offset patterns;
- operator training to recognise alarm systems with low reliability.

Acknowledgments

Experiment 1 was sponsored by the Naval Air Warfare Center, Orlando, Florida. The opinions expressed are those of the author alone.

References

American National Standards Institute (ANSI) (1991) 'Product Safety Signs and Labels', ANSI 535.4-1991.

Baber, C. (1994) 'Psychological Aspects of Conventional In-car Warning Devices', in N.A. Stanton (ed.), *Human Factors in Alarm Design*, London: Taylor & Francis.

Barber, B. (1993) *The Logic and Limits of Trust*, New Brunswick, NJ: Rutgers University Press.

Benson, A.J. and Gedye, J.L. (1963) 'Logical processes in the resolution of orienting conflict' (Final Report 259), Farnborough: Royal Air Force Institute of Aviation Medicine.

Berson, B.L., Po-Chedley, D.A., Boucek, G.P., Hanson, D.C., Leffler, M.F. and Wasson, R.L. (1981) 'Aircraft alerting systems standardization study; Volume II – aircraft alerting system design guidelines' (FAA Report FAA-RD-81-38-II), Washington, DC: Federal Aviation Administration.

Billings, C.E. (1991) 'Human-centered aircraft automation: A concept and guidelines' (Technical Memorandum 103885), Moffett Field, CA: National Aeronautics and Space Administration.

Bliss, J.P. (1989) 'The APTS battery as a surrogate measure of tank gunnery simulator performance', unpublished master's thesis, University of Central Florida, Orlando.

Bliss, J.P. (1993) 'The cry-wolf phenomenon and its effect on operator responses', unpublished doctoral dissertation, University of Central Florida, Orlando.

Bliss, J.P. (1997) 'Alarm reaction patterns by pilots as a function of reaction modality', *International Journal of Aviation Psychology*, 7(1), 1–14.

Bliss, J.P. (in preparation) 'A review of alarm failure in civilian air accidents', Draft manuscript.

Bliss, J.P. and Gilson, R.D. (1998) 'Emergency signal failure: Implications and recommendations', *Ergonomics*, 41(1), 57–72.

Bliss, J.P., Dunn, M. and Fuller, B.S. (1995) 'Reversal of the cry-wolf effect: An investigation of two methods to increase alarm response rates', *Perceptual and Motor Skills*, 80, 1231–42.

Bliss, J.P., Jeans, S.M. and Prioux, H.J. (1996) 'Dual-task performance as a function of individual alarm validity and alarm system reliability information', *Proceedings of the Human Factors and Ergonomics Society, 40th Annual Meeting*, Santa Monica, CA: Human Factors and Ergonomics Society.

Breznitz, S. (1983) *Cry Wolf: The psychology of false alarms*, Hillsdale, NJ: Lawrence Erlbaum Associates.

Cardosi, K.M. and Boole, P.W. (1991) 'Analysis of pilot response time to time-critical air traffic control calls' (Final Report No. DOT/FAA/RD-91/20), Washington, DC: Federal Aviation Administration.

Comstock, J.R. and Arnegard, R.J. (1992) 'The multi-attribute task battery for human operator workload and strategic behavior research' (Technical Memorandum No. 104174), Hampton, VA: National Aeronautics and Space Administration.

Damos, D.L. (1985) 'The effect of asymmetric transfer and speech technology on dual-task performance', *Human Factors*, 27(4), 409–21.

Duchon, J.C. and Laage, L.W. (1986) 'The consideration of human factors in the design of a backing-up warning system', *Proceedings of the 30th Annual Meeting of the Human Factors Society*, Santa Monica, CA: Human Factors Society.

Dunn, M.C. (1995) 'Primary-task and secondary-task workload and the cry-wolf phenomenon', unpublished doctoral dissertation, The University of Alabama in Huntsville.

Edwards, W. (1987) 'Decision making', in G. Salvendy (ed.), *Handbook of Human Factors*, New York: Wiley.

Edworthy, J. (1994) 'Urgency mapping in auditory warning signals', in N.A. Stanton (ed.), *Human Factors in Alarm Design*, London: Taylor & Francis.

Edworthy, J. and Adams, A. (1996) *Warning Design: A Research Prospective*, London: Taylor & Francis.

Edworthy, J., Hellier, E. and Hards, R. (1995) 'The semantic associations of acoustic parameters commonly used in the design of auditory information and warning signals', *Ergonomics*, 38(11), 2341–61.

Edworthy, J., Loxley, S. and Dennis, I. (1991) 'Improving auditory warning design: Relationship between warning sound parameters and perceived urgency', *Human Factors*, 33(2), 205–32.

Getty, D.J., Swets, J.A., Pickett, R.M. and Gonthier, D. (1995) 'System operator response to warnings of danger: A laboratory investigation of the effects of the predictive value of a warning on human response time', *Journal of Experimental Psychology: Applied*, 1, 19–33.

Gilson, R.D. and Phillips, M. (1996) 'Alerts or sound alarm?', *Aerospace Engineering*, May, 21–3.

Gilson, R.D., Deaton, J.E. and Mouloua, M. (1996) 'Coping with complex alarms', *Ergonomics in Design*, October, 12–18.

Green, D.M. and Swets, J.A. (1966) *Signal Detection Theory and Psychophysics*, New York: Wiley.

Haas, E. and Edworthy, J. (1996) 'Measuring perceived urgency to create safe auditory warnings', *Proceedings of the Human Factors and Ergonomics Society, 40th Annual Meeting*, Santa Monica, CA: Human Factors and Ergonomics Society.

Hellier, E., Edworthy, J. and Dennis, I. (1993) 'Improving auditory warning design: Quantifying and predicting the effects of different warning parameters on perceived urgency', *Human Factors*, 35(4), 693–706.

Herrnstein, R.J. (1961) 'Relative and absolute strength of response as a function of frequency of reinforcement', *Journal of the Experimental Analysis of Behavior*, 4, 267–72.

Hopkin, V.D. (1995) 'The impact of automation on air traffic control systems', in J.A. Wise, V.D. Hopkin and M.L. Smith (eds), *Automation and systems issues in air traffic control*, Berlin: Springer-Verlag.

Jones, M.B. (1979) 'Stabilization and task definition in a performance test battery', (Final Report NBDL-M001), New Orleans, LA: Naval Biodynamics Laboratory.

Kerr, J.L. (1985) 'Auditory warnings in intensive care units and operating theatres', *Ergonomics International*, 85, 172–4.

Kerstholt, J.H., Passenier, P.O., Houttisin, K. and Schuffel, H. (1996) 'The effect of a priori probability and complexity on decision making in a supervisory control task', *Human Factors*, 38(1), 65–78.

Lee, J. and Moray, N. (1992) 'Trust, control strategies, and allocation of function in human–machine systems', *Ergonomics*, 35(10), 1243–70.

Mallett, L.G., Vaught, C. and Brnich, M.J. (1992) 'Sociotechnical communication in an underground mine fire: A study of warning messages during an emergency evacuation', *Safety Science*, 16, 709–28.

Marshall, E. and Baker, S. (1994) 'Alarms in Nuclear Power Plant Control Rooms: Current Approaches and Future Design', in N.A. Stanton (ed.), *Human Factors in Alarm Design*, London: Taylor & Francis.

Mastroianni, G.R. and Schopper, A.W. (1986) 'Degradation of force-loaded pursuit tracking performance in a dual-task paradigm', *Ergonomics*, 29(5), 639–47.

McDonald, D.P., Gilson, R.D., Mouloua, M. and Deaton, J.E. (1995) 'The effect of collateral alarms on primary response behavior', *Proceedings of the Human Factors and Ergonomics Society, 39th Annual Meeting*, Santa Monica, CA: Human Factors and Ergonomics Society.

McIntyre, J.W.R. (1985) 'Ergonomics: Anaesthetists' use of auditory alarms in the operating room', *International Journal of Clinical Monitoring and Computing*, 2, 47–55.

Meredith, C. and Edworthy, J. (1994) 'Sources of confusion in intensive therapy unit alarms', in N.A. Stanton (ed.), *Human Factors in Alarm Design*, London: Taylor & Francis.

Moray, N. (1981) 'The role of attention in the detection of errors and the diagnosis of errors in man–machine systems', in J. Rasmussen and W. Rouse (eds), *Human Detection and Diagnosis of System Failures*, New York: Plenum Press.

Muir, B.M. (1994) 'Trust in automation: Part I. Theoretical issues in the study of trust and human intervention in automated systems', *Ergonomics*, 37(11), 1905–22.

Paté-Cornell, M.E. (1986) 'Warning systems in risk management', *Risk Analysis*, 6(2), 223–34.

Patterson, R.D. (1982) 'Guidelines for Auditory Warning Systems on Civil Aircraft', CAA paper 82017, London: Civil Aviation Authority.

Putz, V.R. and Rothe, R. (1974) 'Peripheral signal detection and concurrent compensatory tracking', *Journal of Motor Behavior*, 6(3), 155–63.

Rempel, J.K., Holmes, J.G. and Zanna, M.P. (1985) 'Trust in close relationships', *Journal of Personality and Social Psychology*, 49, 95–112.

Robinson, G.H. (1991) 'Partial attention in warning failure: Observations from accidents', *Proceedings of the Human Factors Society, 35th Annual Meeting*, Santa Monica, CA: Human Factors Society.

Rodriguez, M.A. (1991) 'What makes a warning label salient?', *Proceedings of the*

Human Factors Society, 35th Annual Meeting, Santa Monica, CA: Human Factors Society.

Sanders, M.S. and McCormick, E.J. (1993) *Human Factors in Engineering and Design*, New York: McGraw-Hill.

Selcon, S.J., Taylor, R.M. and McKenna, F.P. (1995) 'Integrating multiple information sources: Using redundancy in the design of warnings', *Ergonomics*, **38**(11), 2362–71.

Seminara, J.L., Gonzalez, W.R. and Parsons, S.O. (1977) 'Human factors review of nuclear power plant control room design', Electric Power Research Institute Report NP-309.

Sorkin, R.D. (1988) 'Why are people turning off our alarms?', *Journal of the Acoustical Society of America*, **84**(3), 1107–8.

Sorkin, R.D. and Woods, D.D. (1985) 'Systems with human monitors: A signal detection analysis', *Human–Computer Interaction*, **1**, 49–75.

Sorkin, R.D., Kantowitz, B.H. and Kantowitz, S.C. (1988) 'Likelihood alarm displays', *Human Factors*, **30**(4), 445–59.

Stanton, N.A. (1994) *Human Factors in Alarm Design*, London: Taylor & Francis.

Stanton, N.A. and Baber, C. (1995) 'Alarm-initiated activities: An analysis of alarm handling by operators using text-based alarm systems in supervisory control systems', *Ergonomics*, **38**(11), 2414–31.

Stevens, S.S. (1957) 'On the psychophysical law', *Psychological Review*, **64**(3), 153–80.

Turnage, J.J. and Kennedy, R.S. (1992) 'The development and use of a computerized human performance test battery for repeated-measures applications', *Human Performance*, **5**(4), 265–301.

Turnage, J.J., Kennedy, R.S., Gilson, R.D., Bliss, J.P. and Nolan, M.D. (1988) *The use of surrogate measurement for the prediction of flight training performances* (IST Faculty Grant Award), Orlando, FL: University of Central Florida Institute for Simulation and Training.

Tversky, A. and Kahneman, D. (1981) 'The framing of decisions and the psychology of choice', *Science*, **211**, 453–8.

Tyler, R.R., Shilling, R.D. and Gilson, R.D. (1995) 'False alarms in naval aircraft: a review of naval safety center mishap data' (Special Report 95-003), Orlando, FL: Naval Air Warfare Center Training Systems Division.

Wickens, C.D. (1984) 'Processing resources in attention', in R. Parasuraman and R. Davies (eds), *Varieties of Attention*, New York: Academic Press.

Wickens, C.D. (1992) *Engineering Psychology and Human Performance*, (2nd edn), New York: Harper-Collins.

Wogalter, M.S., Godfrey, S.S., Fontenelle, G.A., Desaulniers, D.R., Rothstein, P.R. and Laughery, K.R. (1987) 'Effectiveness of warnings', *Human Factors*, **29**(5), 599–612.

10 Using Redundancy in the Design of Time-critical Warnings: A Theory-driven Approach

STEPHEN J. SELCON, *Defence Evaluation and Research Agency*

Introduction

As the time pressure and information overload on pilots increases, the need for appropriately designed warnings becomes ever more important. The function of such warnings is twofold. The first is to alert the pilot to the fact that a potentially hazardous situation has occurred and to direct his or her attention towards the problem. This needs to be achieved against a background of high ambient noise and, often, excessive cognitive and perceptual workload. Of equal importance is to gain the pilot's attention without producing a startle response. Failure to gain attention will result in warnings being missed or ignored. Startling the pilot will result in overarousal, which will interfere with the operator's ability to respond effectively to the situation. Potential human factors solutions to these requirements are discussed in detail elsewhere in this book.

The second function of an effective warning is to inform the operator, as rapidly and unambiguously as possible, of the nature of the hazardous situation being represented. It is this second aspect that will be addressed in this chapter. The discussion here will assume that attention has been gained and that attentional priority will be given to the warning. For high-priority warnings in particular, the time available for a remedial response to be made can be as little as one or two seconds. If attentional priority is not given to that warning, the aircraft is likely to be lost. Therefore it seems reasonable to assume that, once alerting has occurred, the pilot will focus on such warnings to the temporary exclusion of other less time- and safety-critical cockpit tasks. An example would be in avoiding incoming missiles and deploying appropriate countermeasures. Failure to perform such manoeuvres adequately will result in the aircraft being shot down, thus making all other concurrent cockpit tasks effectively irrelevant until the threat is defeated.

For high-priority warnings, the reduction of response times through effective human-centred design of the warning stimuli will have significant benefits. In such situations, savings of even a few milliseconds can, on occasion, literally prove the difference between life and death. This chapter will therefore provide a summary from the literature of one such method of reducing decision times, namely the principle of redundancy gains. The ability to represent multiple concurrent forms of the same information has been greatly increased by the availability of auditory displays in the cockpit. Such displays now include digitised speech and three-dimensional spatial auditory representations of threats. This technology allows the designer of warnings much greater scope and flexibility in using redundant information presentation to improve the informing properties of the warning stimuli. Empirical demonstrations of ways in which redundancy gain theory can be implemented successfully in the design of aircraft warnings will be given.

It should also be remembered that, in informing the user about the nature of a warning situation, it is not only speed of response that is important. It can be argued that the depth of understanding produced is potentially as critical as the speed with which it is gained. This is because the operator must respond not only quickly, but also correctly. To this end, the underlying psychological mechanisms of the redundancy gain phenomenon will be considered. Potential implications for depth of understanding, and consequently situational awareness, will be drawn from neural models of the redundancy gain effect. The impact of such processing models for the design of safety critical warning systems will be shown.

Redundancy Gains

The traditional concept of 'redundancy gains' referred to the facilitation of target recognition performance through the simultaneous presentation of multiple, identical targets (see Eriksen *et al.*, 1989, for an overview). More specifically, Grice and Canham (1990) provide the following definition of a redundancy gain: 'an improvement in accuracy in a recognition experiment or a decrease in latency in a reaction time experiment when the target is presented more than once'.

A typical example of the experimental paradigms used to investigate redundancy gain effects is given by Grice and Gwynne (1987). Two targets (for example, the upper-case letters H and S) were presented, with the subject required to categorise whether the target displayed was an H or an S. Choice reaction times (CRT) were recorded. In half of the trials, subjects were presented with two identical target letters simultaneously. In the other half, one target was replaced by an irrelevant noise letter (for example, the upper-case letter Y). When two targets were presented simultaneously (for example, HH or SS), response times were lower than when a single presentation of the target is given (HY, YH, SY or YS). The presence of the second target was redundant (since a single target was sufficient to complete the task). Thus the facilitation of performance from the redundant target is referred to as a redundancy gain.

Two theoretical explanations of this phenomenon have been advanced. First, Raab (1962) proposed a separate-activation model with the two targets entering an 'attentional-race'. The target which wins the attentional race will be the one which is responded to. If the time to respond to each target varies across trials, then the average finishing time for the faster of two parallel processes will be shorter than the average finishing time for either process alone. In other words, the laws of probability will mean that always taking the faster of two target responses will necessarily be faster, over a number of trials, than responses to a single target. Miller (1982) specified this theoretical position in terms of a mathematical formula, prescribing the necessary statistical parameters which would need to be met by the experimental data, known as the 'race model inequality'. The inequality specifies that the probability of trials (in terms of the cumulative distribution function) containing two stimuli being responded to faster than time t cannot be greater than the probability of each stimulus in isolation being responded to faster than time t. In other words, since the attentional race model is based on taking whichever target can be responded to most quickly, the total distribution of trials with both stimuli cannot contain more fast responses than the sum of the two targets in isolation. Any violation of this statistical expression, says Miller, can be taken as evidence that the attentional race model is insufficient to explain the redundancy gain effect.

Miller (1982) experimentally investigated whether such a violation of the race model inequality could be observed. Miller used letter-pair stimuli, where each letter in the pair could be either the target (the letter A) or noise (the letter B). Thus three types of letter pair were used, redundant target (AA), single target (AB or BA) and no target (BB). Subjects were required to respond as to whether a target was present or not, using a two-button response box marked 'Yes' and

'No'. Miller found shorter mean RTs for redundant targets (that is, when the stimulus pair was AA) with significant violations of the statistical inequality. In light of these findings, Miller proposed an alternative explanation of the redundancy gain effect, based on a coactivation model of the processing occurring. The coactivation model proposes that the two targets 'coactivate' processing channels corresponding to a response, rather than racing to trigger it. Two information sources will provide more activation to that response, in a given time, thus allowing a decision on the selection of that response to be achieved more rapidly.

Miller's findings have been criticised, however, by van der Heidjen *et al.* (1984) and by Grice *et al.* (1984). Both of these papers suggest that the Miller (1982) results can equally be explained by response competition from the noise stimulus interfering with processing of the target, as by the redundant second target facilitating performance. In other words, it may be that the presence of the noise stimulus (the letter B) was slowing subjects in the single target condition, rather than the redundant secondary target speeding performance in the redundant target condition. Such an interpretation would make it impossible to draw conclusions about the race model inequality and the coactivation model, since it is not clear whether it is indeed a redundancy gain effect producing the differences.

The issue of whether performance differences were as a result of redundancy gains or response competition was addressed by Grice and Canham (1990). They replicated the method used by Grice and Gwynne (1987), with the exception that they used a go/no go RT task, instead of a CRT task. In this task, subjects were required to respond only if they were presented with one or more targets. They presented the single target condition both with and without the noise letter Y. They found that redundancy gains could be observed with and without the noise letter present. Further the magnitude of the gain found was not affected by the presence/absence of the noise stimulus. Thus if two presentations of a letter are responded to more quickly than that letter in isolation (that is, with no noise letter) the effect can be said to be a genuine redundancy gain.

Further evidence of redundancy gains with a go/no go task were demonstrated by Grice and Reed (1992). They used redundant targets which were not physically identical (such as A and a), presented as both a CRT and a go/no go task. They found no redundancy gain when compared to a single target with no noise for the CRT task. For the go/no go task, however, using the same stimuli, a significant redundancy gain was observed. This implied that the redundancy gain was not the product of response competition, because the two target condition was being compared to a single target in isolation. In a second experiment, Grice and Reed (1992) used redundant targets which were neither physically nor linguistically identical (for example the letters A and D). They again found a true redundancy gain with the go/no go task, but not with the CRT task. They also found, in both experiments, a significant violation of the race model inequality. This can be taken as support for the rejection of the race model as argued by Miller (1982). A similar violation of the race model inequality demonstrated by Mordkoff and Miller (1993), who replicated the results obtained by Grice and Reed (1992) with the exception of balancing for biased contingencies (that is, the number of times a given type of trial occurred). They again found both redundancy gains and violations of the race model inequality, providing further evidence against the race model proposed by Raab (1962).

Thus, although to date the number of studies examining redundancy gains is relatively small, sufficient evidence has been found to demonstrate that redundancy gains (which are not an artefact of response competition) from two versus one presentation of a target can be obtained, albeit that such effects appear to be task-dependent. A tentative model, namely coactivation, of the cognitive processes producing such effects has been advanced, and limited support for it found. Such support, to date, has been more to reject the race model theory than to validate the coactivation model. Further (and more direct) testing of the coactivation model will be required before firm conclusions can be drawn from it. The concept of redundancy gains does, however, provide a useful starting point for considering how more than one source of information can be integrated to facilitate performance.

The concept of redundancy gains, described above, is a useful one. Although the authors cited here have restricted its use to the example of multiple representations of targets (albeit that the targets can differ from each other), it might also be applied to multiple dimensions of stimuli. To this end, the next section will examine the idea of performance facilitation and interference being produced by covarying dimensions of a single stimulus, namely the concept of 'Perceptual Integrality'.

It is also worth addressing the applicability of the redundancy gain concept to the design of aircraft cockpit systems. Any psychological effect which is known to increase accuracy and/or reduce response latencies is of potential advantage to the designer of time- and safety-critical systems. The redundancy gain effects described here imply that the benefits are not necessarily limited to the simplistic attentional competition effects for which redundancy has traditionally been used. The implication of the experimental results described above, particularly within the context of a coactivation model, is that speed and accuracy improvements can be achieved when both sources are available for processing concurrently, and that they may be produced through the integration of information during processing.

Perceptual Integrality

The concept of multidimensional stimuli and the way they are processed was addressed by Garner and Felfoldy (1970). They examined subjects' ability to sort stimuli varying concurrently on one or two dimensions of colour, using the Munsell dimensions of colour value and chroma. They measured subjects' ability to sort the cards using one dimension while a second dimension was varied in a correlated manner or orthogonally (within a single stimulus), and compared it with performance when only one dimension varied and the second was kept constant. The stimuli were found to produce facilitation when the information on the different dimensions was varied in correlated manner, and interference when it was varied in an orthogonal manner. Facilitation was found when the correlation was either positive (high value/high chroma, low value/low chroma) or negative (high value/low chroma, low value/high chroma). They explained this effect in terms of the dimensions being perceptually integral: that is, the effect of one dimension could not be separated from the effect of the other. They replicated the experiment using separate stimuli for each dimension (with the two stimuli displayed concurrently in adjacent spatial positions). This time they found neither facilitation (from the correlated dimensions) nor interference (from the orthogonal dimensions). This they interpreted as the stimulus dimensions being separable (in this case owing to their spatial separation into discrete stimuli, rather than multidimensional aspects of the same stimulus). Such a finding, they claimed, demonstrated the concept of perceptual integrality; namely, that 'the structure of the perceived relations between multi-dimensional stimuli depends on whether the stimulus dimensions are integral or separable, the distinction phenomenologically being between dimensions which can be pulled apart, seen as unrelated, unanalysable, and those which cannot be analysed but somehow are perceived as single dimensions' (Garner and Felfoldy, 1970, p.225).

The effects described so far have all shown stimulus dimension interactions within the visual modality. Melara and O'Brien (1987) investigated whether integrality effects could be achieved across, as well as within, sensory modalities. The stimuli they investigated were the visual dimension of vertical position and the auditory dimension of pitch (frequency). These dimensions are 'synaesthetically corresponding' in that they are perceived as being related perceptually. Marks (1975) found that subjects reported high pitch sounds as corresponding to high spatial position, and low pitch sounds to low spatial positions. Melara and O'Brien found an interference effect when stimulus dimensions were orthogonal (that is, the second dimension varied independently of the first) and a facilitation effect when stimulus dimensions were positively correlated (high position/high pitch and low position/low pitch). A similar pattern of results was found by Melara (1989) using auditory pitch and visual colour dimensions. Melara expressed concern that, since the stimuli used by Melara and O'Brien had common linguistic labels, any facilitation found might be occurring as a result of linguistic rather than perceptual processes. Colour and pitch are also synaesthetically corresponding dimensions (low pitch/black and high pitch/white), but do not have any semantic or linguistic commonality. Melara was able to produce interference from orthogonal, and facilitation from positively correlated, stimulus dimension combinations. The implication he drew from the study was that the observed effects were occurring because of dimensional integrality during stimulus perception. Thus redundancy gains can be achieved from presenting cross-modal information as representations of a common stimulus dimension.

The concepts of integral/separable stimuli have been used in the design of information on visual displays. Jacob *et al.* (1976) used 'integral' displays to improve performance using graphical information. Goettl *et al.* (1986) compared integral and separable graphs for multi-cue judgments. Carswell and Wickens (1990) review such applications of the integrality paradigm in information design and conclude that 'integral displays are related to superior performance only when subjects must process each of the displayed variables in order to make a single response. Separable displays seem to better support tasks requiring information filtering, localisation, or performance of several independent tasks simultaneously'. Such a design principle is in line with the predictions made by the theory of integrality/separability of stimuli. The concept of performance facilitation through the interaction of correlated stimulus dimensions has also been applied to the design of aviation-specific visual displays. Farmer *et al.* (1980) used the concept of integral dimensions to predict colour difference requirements, on the three Munsell colour dimensions, for cockpit map design. Taylor (1984) used combinations of stimulus dimensions, providing integral cues, in the design of Head Up Display (HUD) symbology. By evaluating stimulus dimension integrality, with both correlated and orthogonal dimension combinations, Taylor was able to demonstrate improved aircraft attitude awareness through the use of displays composed of integral stimulus dimensions. Thus applied benefits can be gained through the implementation of concepts such as perceptual integrality in design.

The concept of perceptual integrality, as discussed here, is considered to be available during the perceptual encoding and organisation of information. No consideration has been given to the potential for a cognitive analogue of the integrality effect to exist. Indeed Garner and Felfoldy (1970) specifically describe the effect as being dependent on the composition of the multidimensional perceptual stimulus rather than the cognitive integration of information. The ability to produce facilitation and interference effects at a cognitive level (that is, where dimensions are becoming integrated at more advanced stages of processing than perceptual encoding) is addressed in the remainder of this chapter.

The essence of the perceptual integrality results discussed above is that the presentation of more than one congruent stimulus dimension can produce facilitation of performance. Since the tasks involved can be carried out using a single stimulus dimension, the second dimension can be said to be redundant (in much the same way as multiple targets are redundant). Since they facilitate performance they can literally, at least, be described as redundancy gains. There is value in considering the degree of overlap between the two sets of findings since the redundancy gain target experiments are concerned with the integration of the redundant information during processing, rather than at the level of the perceptual encoding as described by the perceptual integrality studies. This allows consideration of the integration of multiple information sources at the cognitive level, as an analogy of the perceptual integrality/separability effects (Melara and Mounts, 1993). A major source of information on the cognitive integration of information from different sources can be found in the literature on Stroop effects which is discussed in the next section.

The Stroop Effect

Stroop (1935) demonstrated that the presence of an incongruent secondary source of information interfered with the processing of the primary information, producing increased task completion times. This result has led to a mass of experimental investigations into the effect. A comprehensive review of these is given by McLeod (1991). This chapter will focus on those effects which have potential relevance to the applied aviation environment. The original Stroop paradigm used colour-naming words presented in inappropriate ink colours. The method of presentation used was by large cards, on which approximately a hundred stimuli were written. Subjects were required to name the ink colour aloud (a verbal response) while ignoring the colour-naming word which was written in that ink. The time to name all the stimuli on the card was taken as the performance measure. The presence of the incongruent colour-naming words produced a marked interference in task performance, by increasing the time required to complete the task, as compared to naming the ink colour alone. The converse or 'reverse' Stroop (when the task was to read the colour word when written in an incongruent ink colour) was not found. The absence of the reverse Stroop effect became a characteristic of early research into Stroop interference.

Alternative implementations of the Stroop paradigm continued to show the same interference effect. Dalrymple-Alford and Budyar (1966) first used individual stimulus, rather than word list, presentation of Stroop stimuli to show the same interference pattern as that found with the original paradigm. The individual stimulus technique continued to use a verbal response. A markedly different approach was developed by Tecce and Happ (1964). They used a card-sorting task, where stimuli were written on individual cards, with subjects being required to sort the cards into a number of piles corresponding to the ink colour in which words were written on them. Thus the response modality was now manual rather than verbal. A similar pattern of interference was observed, although other researchers have shown that the interference achieved, although being correlated with the original effect, was reduced in magnitude (Taylor and Clive, 1983). More recently the advent of microcomputers has allowed the mechanisation of the above implementations, thus further increasing the utility of the paradigm.

It can be seen from the above discussion that the Stroop effect is a robust and replicable paradigm. It is this robustness, even with quite different implementations of the task, that has led to its widespread usage. The ability to alter the method of presentation of stimuli, and the response to them, has allowed the paradigm to be used to investigate a much wider range of issues than merely the interference between ink colours and words. Some of these are discussed in the next sections.

Reverse Stroop Effects

The reverse Stroop effect is said to occur when the presence of an incongruent ink colour interferes with the naming of the colour word written in that ink. It was first demonstrated by Chmiel (1984) using the card-sorting task. He found that, when the task required of subjects was to sort cards by colour word into bins identified by ink colour patches, the presence of incongruent ink colours in which the colour words were written produced an interference effect. When the bins were identified by colour words, the reverse Stroop was not evident (as in Stroop's original experiments). More recent research has also found that an analogy to the reverse Stroop can be obtained (Dunbar and McLeod, 1984; McLeod and Dunbar, 1988; Cohen *et al.*, 1990) with training. The McLeod and Dunbar (1988) experiments used an analogue of the Stroop task using shapes (given an arbitrary assignment to colour names) and ink colours. They found that the pattern of interference obtained was dependent on the degree of practice on the task. With a small amount of practice on the shape–colour name association, ink colour interfered with shape naming but not the converse. With more practice, however, the amounts of interference in each direction became equivalent. With a high degree of practice, the ink colour failed to produce interference to the shape–colour naming, whilst the shapes produce a high degree of interference to ink colour naming. Thus the direction of Stroop interference is in no way fixed, and is dependent on both the type of task used and the nature and familiarity of the stimuli. The implication of such a finding is that many more types of stimuli are potentially able to demonstrate the Stroop effect than Stroop himself might have originally envisaged, thus increasing its potential relevance for implementation in aircrew systems design.

Picture–Word Analogues

The nature of the stimuli used in the original Stroop task allows only limited variation of the stimuli for experimental manipulation. Because of this, a popular alternative has been to use picture–word combination tasks to investigate the effect. The first use of picture–word combinations was by Hentschel (1973), who embedded incongruent words within line-drawings and required subjects to name the pictures. He found that the words interfered with picture naming in a manner analogous to the Stroop effect. Rosinski *et al.* (1975) showed, using a similar task, that the reverse Stroop (that is, pictures interfering with words) showed no effect, thus supporting the analogy between the colour and picture–word paradigms.

The ability to use pictures, rather than colours, allowed experimental manipulation of semantic relations between stimuli to study the effect of semantic information on interference. Rosinski (1977) showed that commonality of semantic category between picture–word pairings affected the degree of interference obtained, with incongruent words of the same semantic category as the picture producing greater interference than words from different semantic categories (for example, the word 'ankle' on the picture of a hand). Lupker (1979), however, found that picture–word pairings with common semantic associations, but from different semantic categories (for example, the word 'cheese' on a picture of a mouse), failed to show increased interference as compared to semantically unrelated words. This issue has been clarified by La Heij *et al.* (1990), who demonstrated that interference and facilitation can occur with both categorically and associatively linked picture–word combinations, with the extent of the interference/facilitation being dependent on the interaction between the type of semantic link and the stimulus onset asynchrony between the individual parts of the paired stimulus.

These findings have particular relevance for the design of information in military cockpits since they demonstrate that the presentation of related information, even when separated in time, can affect the way such information is processed. Further, such interaction between information is possible even when the coding of the two types of information is different: that is, pictorial versus linguistic coding. In the cockpit, where large quantities of differing information are concurrently available, consideration must be given to the design of that information to prevent interference occurring.

Auditory Analogues

Although consideration has only been give thus far to interference effects found within the visual modality, auditory analogues of the Stroop effect have also been demonstrated. Simon *et al.* (1975) showed that, for pairs of incongruent auditory stimuli, Stroop interference effects could be produced even when the presentation of the two information sources was separated in time by up to 400 milliseconds. Hamers and Lambert (1972) showed that, when the spoken words 'high' and 'low' were presented at high (175Hz) and low (110Hz) pitches, word identification performance was worse when the pitch–word combination was incongruent. They took this as evidence that interference was being caused by the incongruent pitch information. However, as there was no neutral pitch control, it is impossible to separate interference between the incongruent pairs from the facilitation from the congruent pairs. This concept of facilitation from congruent information will be discussed in detail later in this chapter. Zakay and Glickson (1985) showed a similar effect with musical notes presented with incongruent musical notation. Further they showed that response modality, and stimulus–response compatibility, affected the interference obtained. This is in line with the findings of McClain (1983) who replicated the Hamers and Lambert task using three response modes: verbal, button press and pitched hum. McClain found that interference only occurred when the stimulus dimension to be reported and the response modality differed (pitch–verbal, pitch–button and word–hum) and not when they were compatible. This stimulus–response link has been postulated as an explanation for reverse Stroops not being obtained with the original Stroop paradigm (McLeod, 1991). Since that paradigm uses a verbal response, there is strong stimulus–response compatibility between the word stimuli and the response modality. The colours, however, have no such intuitive compatibility and hence are more open to interference by the secondary information source.

The ability to produce auditory Stroop effects is also relevant to the aviation context. Although the pilot's primary information source is visual, an increasing amount of auditory information is becoming available. This includes dialogue with other crew members, radio communications and tone or voice warning systems. Thus the way in which auditory information interacts both with other auditory information and with visual information needs to be taken into account in designing cockpit displays.

Cross-modality Stroop Effects

Although less research has been published on the topic, there is evidence that Stroop-type effects can be produced by stimuli across, as well as within, sensory modalities. A cross-modal interference effect was demonstrated by McGurk and MacDonald (1976) and MacDonald and McGurk (1978). They presented contradictory phonemic and articulatory information simultaneously to subjects and found that inappropriate articulation (that is, uncorrelated speaker lip movements) changed the phoneme which was perceived. An example would be the subject seeing an articulation of the phoneme/ga-ga/ while hearing the phoneme/ba-ba/. The resultant perception in 98 per cent of the adult subjects tested was the phoneme/da-da/. This implies that information is being integrated across the two modalities, thus affecting the resultant perception: that is, the incongruence of the information sources is interfering with the perception of either independent source. Although the authors themselves do not claim this as a Stroop effect, the analogy to the effects described above seems clear. Simon and Craft (1970) presented an auditory tone to either the left or right ears at the same time as subjects were required to respond to the position of either a left or right light. They found that responses were faster when the tone was presented to the congruent ear than to the incongruent ear. Again this experiment fails to clarify whether the effect was the result of interference from the incongruent stimuli or facilitation from the congruent stimuli. It does, however, provide further evidence for the concurrent integration of information from different modalities to produce performance changes analogous to the Stroop effect.

As with auditory Stroop analogues, there are potential implications for aviation system design. Since the pilot is likely to be attending to both auditory and visual information, often simultaneously, the possibility of interaction between the information sources must be considered. This is particularly the case when considering visual point-of-fixation displays, such as Helmet Mounted Displays (HMDs), being used in conjunction with increasing numbers of auditory displays, such as voice warnings and spatial (3-D sound) auditory displays. Further the potential to provide congruent information simultaneously to two or more modalities may provide facilitatory effects which can be specifically designed into information displays.

Stroop Facilitation

The previous sections have concentrated on the classic Stroop interference effects and their analogues. Research has shown, however, that a 'positive' Stroop effect can produce facilitation when the secondary source of information is congruent with the primary source (Dyer, 1971, 1973). Such facilitation has always been much less, however, than the equivalent interference (Glaser & Glaser, 1982). Such facilitation has also been shown for Stroop analogues. Ehri (1976) demonstrated facilitation in a picture–word task using the picture only as a control. Glaser and Dungelhoff (1984) showed a similar effect using a row of Xs as a control, matched in size and spatial position to the words in the experimental condition. Both of these studies also showed that the amount of facilitation produced by congruent stimuli was, as in the colour–word studies, considerably less than the amount of interference produced by incongruent stimuli. In a cross-modal Stroop analogue, MacLeod and Summerfield (1987) showed that, where correlated facial articulatory movements were presented with vocalisations, identification of words within sentences was improved. This performance gain was equivalent to that achieved with a 10 to 12dB increase in the signal-to-noise ratio.

Thus facilitation can be achieved across and within modalities, and with different types of information coding. Such findings have more direct relevance to the design of aircraft information displays than the findings regarding interference effects. The ability to shorten response times by providing a secondary, congruent source of information has potential benefits in the design of high-priority cockpit systems where speed of response is critical. An example of the application of facilitation effects to the design of cockpit warning systems will be considered in the final section of this chapter. Before such effects can be applied effectively in design, however, an understanding of the cognitive mechanisms underlying them is required. By having an accurate model of the way the information is being integrated to produce facilitation, the designer will be better able to tailor cockpit information to maximise the performance benefits available from concurrent, congruent information.

Models of the Stroop Effect

Several theoretical interpretations of the Stroop effect have been made. McLeod (1991, p.203) listed 'eighteen major empirical results that must be explained by any successful account of the Stroop effect'. He used these as criteria for testing models of the Stroop effect. Examples of the criteria include interference, facilitation, reverse Stroop, practice and stimulus onset asynchrony. McLeod's taxonomy will be used to examine the differing theories of the Stroop effect described here.

Relative Speed of Processing

In its simplest form, this model describes the two competing sources of information as taking part in a 'horse race' (in a manner similar to the attentional race models used to explain redundancy gains earlier). Each is processed in parallel, with the stimulus which is processed most quickly being available as a response first. In the original Stroop paradigm, the word is processed more quickly and hence is available as a response before the colour. However, since the word is not providing the information on which the response is required to be based, it will compete with the slower 'correct' response based on the colour information. It is this competition, and the resolution of it, that produces the time penalty which is seen with interference. Reverse Stroop effects are not typically seen because the word is processed before the colour, and hence the colour information is not available in time to compete with the word information response. Examples of this type of explanation are Morton (1969) and Morton and Chambers (1973).

Although this model can account for many of the empirical Stroop findings, some findings cannot be accommodated. By presenting compound stimuli with stimulus onset asynchronies, it is possible to delay the processing of the faster stimulus by delaying its presentation. This should, according to this model, cause the previously slower stimulus to achieve a response first, thus reversing the Stroop effect. Experimental results (for example, Glaser and Dungelhoff, 1984) do not, however support this, with no reversal of the interference being found.

Automaticity

This model is based on the assumption that different types of information, because of their degree of learning or familiarity, are processed with different levels of automaticity. This in turn implies that they will produce different levels of attentional demand. In the Stroop case, words are processed more automatically than ink colours. Further this higher level of automaticity for word reading makes it impossible to disregard the verbal information. Thus the word will attempt to produce a response, with consequent interference to the colour-naming task. The converse will not occur, since the low level of automaticity of the ink colour processing is easily ignored in favour of word processing. Examples of such an interpretation come from Posner and Snyder (1975) and Shiffrin and Schneider (1977). These authors describe automaticity as a gradient, not an all-or-none phenomenon. This allowed their model to account for gradations in the amount of interference being obtained in Stroop analogues, and also the effect of training on the interference being obtained.

Although able to explain more of the empirical data than speed of processing models, there are still potential shortcomings with the model. The automaticity model would predict that, as long as the two stimuli were within the foveal field of view, the word should still be processed automatically. Thus there would be difference in the pattern of interference produced. Kahneman and Henik (1981), however, found that spatial separation of the colour and word stimuli by small distances reduced the amount of interference observed. Furthermore attentional cueing as to the probability of a successive trial containing a congruent or incongruent pairing affects the amount of interference found (for example, Kahneman and Chajczyk, 1983). If word processing is truly automatic, attentional allocation should have no effect on it.

Parallel Processing Models

More recent accounts have considered the mechanism of the Stroop effect to be based on a parallel processing model accruing evidence towards a decision (Logan, 1980). In such a model, evidence towards a decision is accrued over time, until a threshold is reached and a response can be made. Such evidence can be gathered from multiple information sources concurrently (that is, in parallel). If the evidence from all dimensions is confirmatory, a decision threshold can be reached with less processing of each individual information source, and a response can be made more quickly (in a similar manner to the coactivation model of redundancy gains discussed earlier). If, however, the evidence from the different sources is contradictory, the threshold will be raised, with more processing of the target stimulus being required. This will produce an increase in response times. The degree to which one type of information interferes with another will be dependent on the rate at which each is processed. These processing rates are dependent on a static weighting based on prior exposure and practice (the more familiar and learnt a stimulus, the greater the weight given to its evidence) and a variable weighting based on strategic attention allocation. Thus a highly learnt stimulus, such as a word, will produce a high degree of interference when it contradicts the target stimulus because it will have a high weighting. A less well learnt stimulus, such as ink colour, will have less effect because of its lower weighting.

A more explanatory representation of the parallel processing model was made by Cohen *et al.* (1990) using the Parallel Distributed Processing (PDP) model of Rumelhardt *et al.* (1986). This account has very many similarities to the Logan (1980) model, but it is able to specify the nature of the parallel processes in more detail and to provide a mathematical model of those processes. The PDP model is based on the concept of neural pathways, formed into interconnecting networks. The intersections of pathways are called nodes. These nodes receive excitatory and inhibitory inputs from other nodes, at the same or adjacent levels in the network, causing their activation level to be altered. For any node, there is an activation threshold which will cause it to 'fire' or send an output to the next nodes in the network. Reaching this activation threshold will be dependent on the number of excitatory and inhibitory inputs received, and also the resting activation level of the node. Resting activation is dependent on prior usage of that node: a regularly used node will have a high resting activation level, thus requiring less activation to reach threshold. The PDP model also contains an attentional component, whereby attention allocation is used to stimulate desired pathways by altering their resting activation (in much the same way as an external stimulus would).

We can apply this model to the Stroop effect. For colour naming, nodes corresponding to the colour will receive excitation from processing of the colour stimulus. They will also, however, receive inhibitory inputs from the word stimulus. Thus the time to activate that node will be increased, owing to the requirement for increased colour processing, as compared to the situation where no contradictory verbal information is present. When the task requires a verbal response it can be assumed that, because colour words are more regularly processed than colour-naming nodes, word nodes will have a higher resting activation level than colour-naming nodes. Thus the inhibitory effect of a contradictory colour stimulus will have relatively little effect on word reading time. This accounts for the absence of the reverse Stroop effect in the standard paradigm. Training, however, will alter the relative resting activation levels of the two pathways, thus allowing the reverse Stroop effect to become evident. The PDP model can also be used to model perceptual integrality. Stimulus dimensions can be said to have become integral when they gain access to a common node/pathway. At such a point, both stimulus dimensions would provide concurrent excitation (for congruent dimensions) or excitation and inhibition (for orthogonal dimensions). Thus the speed at which the pathway would reach its activation threshold would be speeded or slowed, as appropriate. Once activation of the common pathway had occurred, the dimensions could no longer be separated. In other words they had become an integral or unitary stimulus. Separable stimuli would not gain access to a common node and would therefore produce neither facilitation nor interference.

As well as being able to explain the experimental data found in the literature at a conceptual level, Cohen *et al.* (1990) were able to use the PDP model to simulate, and were able to replicate, many of the empirical Stroop effects using a computational network. The findings they simulated include the Stroop asymmetry between the Stroop and reverse Stroop; the fact that that interference exceeds facilitation (Dunbar and McLeod, 1984); stimulus onset asynchrony effects (Glaser and Glaser, 1982); practice effects (McLeod and Dunbar, 1988); and response-set effects (Klein, 1964). In all of these examples, the computational network simulation was able to replicate the empirical findings from those studies. This provides strong support for the PDP model of Stroop effect processing. It should be borne in mind, however, that not all of the 18 criteria set out by McLeod (1991) have as yet been simulated. Successful replication of these other factors, including cross-modal Stroop effects, will be necessary to truly validate the model. Further, since the PDP model has only recently been applied to the interpretation of Stroop effects, there has not yet been time for empirical studies to attempt to define its limitations, as was the case with the speed of processing and automaticity models. Despite this, PDP modelling of Stroop effects does appear to have utility in providing a framework for thinking about information integration, and more particularly with reference to the application of such findings in the context of aircrew systems design.

Applications to Military Aviation Warnings Design

It can be seen from the above review that the performance effects of redundant information are a robust phenomenon which can occur with a large variety of stimuli. This, coupled with the fact that redundancy can produce facilitation, gives it potential utility in the design of information displays for time-critical real-world tasks in dynamic environments. This is particularly relevant in the military aviation environment, where even small time savings can be vital in the interpretation of safety-critical displays such as high-priority, immediate-action cockpit warnings where a response may be required within one to two seconds. In such situations, attentional priority is not an issue because of the critical importance of responding to the warning. Thus any interference to other, less urgent cockpit tasks, arising from the use of multiple modalities for the presentation of redundant information, will not be an issue.

Cockpit information is currently presented primarily on fixed position visual displays. New technology, however, has increased the scope for adding auditory display such as voice warnings, and 3-D auditory displays (sound containing both monaural and binaural spatial cues to allow the localisation of the sound source in three-dimensional space) with possible redundancy gain performance advantages. Also new visual display technologies such as HMDs (which are attached to the pilot's visor and controlled by head/eye position) will allow immediate presentation of visual information to the pilot at the current locus of fixation, thus increasing the opportunity to take advantage of performance facilitation through multimodal information presentation.

The experiments described in the next section attempt to identify any performance advantages that can accrue through the presentation of multiple sources of congruent information in the aircraft cockpit. The experiments also try to ascertain whether cognitive integrality exists and, if so, at what level of processing it occurs. Both the Stroop effect and the perceptual integrality concept of Garner and Felfoldy are used as an attempt to model the cognitive processes which produce these effects. By looking for facilitation from redundant information sources, it is hoped to show that information can be used integrally both within and across modalities in the decision process: that is, that it can be integrated from uni-/multimodal simultaneous presentation, thus enabling a decision on its 'meaning' to be made more quickly. Further the experiments attempt to demonstrate that the level of processing at which the effects occur will depend on the nature of the relationship between the stimuli.

The applied implication of such an effect would be that the

presentation of simultaneous visual and verbal information by cockpit systems would help the user to understand, and respond to, that information more quickly. It is also possible that an increase in depth of understanding and/or situational awareness (SA) can accrue from such integral information. Although two sources of information may become integral during processing, the two pathways used prior to that integration will be separate. Each of the separate pathways will have both shared and discrete links to pathways not directly involved in the task response (for example, semantic associations). Thus the range of associations being produced by the two stimuli is likely to be enriched, with a possible increase in SA.

Although such a model is highly speculative, any improvements in SA which could be produced would have important implications for the design of highly critical aircrew systems, which require rapid situation assessment and accurate, diagnostic reasoning such as warning systems. An example could be for a multimodal fire warning. The two sources of information would not only activate the semantic unit corresponding to fire more rapidly but they will also provide excitation (by the increased number of semantic associations) to a greater number of networks associated to fire. Since many of these associations are likely to be relevant in the comprehension and response to the fire warning, it is possible that SA and/or understanding may be improved.

Experimental Evidence in Support of Redundant Warnings

Although redundancy has been used for many years in warning design, for example a flashing light coupled with alarm tone, there has been little experimental evaluation of the efficacy of such an approach. While this does not in any way argue against such usage, it means that a clear understanding of the best way to implement redundancy, and a quantification of its benefits, is required. In particular, a better understanding of the mechanisms of redundancy gain will potentially allow its benefits to be maximised by knowing what stimulus components are best able to produce facilitation of performance.

An experimental evaluation of using cross-modal information presentation for cockpit systems warnings was conducted by Selcon *et al.* (1992). They used pictorial representations of cockpit warnings coupled with auditory verbal equivalents to investigate two aspects of redundancy gains in a warning categorisation task. First, were such expected gains empirically demonstrable? Second, at what level of processing do such gains occur? The pictorial warning 'icons' were produced by experienced RAF aircrew as being meaningful and recognisable representations of real aircraft warning situations. By using pictorial visual representations and spoken words, Selcon *et al.* ensured that there was no physical or linguistic redundancy between the two information dimensions. Any integration of the information would have to be produced at a semantic level of processing, that is through the combination of the meanings of the stimuli.

The results of the Selcon *et al.* study showed that performance facilitation did indeed occur as a result of the presentation of redundant warnings. Thus the combined visual/auditory condition was significantly faster than either type of warning presented alone. This not only supports the existing use of redundant warnings but also implies that more utility can be obtained from redundancy than is currently being exploited. This is because the study demonstrates that the integration of information is available at relatively deep cognitive levels, as well as between modalities and coding representations. In other words, since redundancy can be achieved through the semantic associations of the stimuli (rather than merely their physical or linguistic properties) and is not limited to only one perceptual modality or coding type, the number of applications for which it can have relevance is increased. Also the number of possible sources of redundant information presentation available to designers is increased since integration of information can occur at levels from initial perceptual encoding (cf. Garner and Felfoldy, 1970) through to truly cognitive integration of stimuli meanings.

In addition to the speed of understanding advantages they found, Selcon *et al.* also found limited support for improved depth of understanding from redundant warning information. They measured situational awareness (SA) using the Situational Awareness Rating Technique (SART) (Taylor, 1989; Selcon and Taylor, 1989) and found some evidence that SA was improved when redundant information was present. Although at a reduced level of significance, possibly because of the lack of sensitivity of the SART technique, the results indicated that subjects had both more SA and better understanding when both types of information were presented. Such a benefit would be of particular consequence in a real-world warning situation where making the appropriate response to the demands of the situation is paramount for survival.

This robustness and flexibility of the redundancy gain phenomenon means that, for a sufficiently high priority warning, where the criticality of the situation justifies maximum design effort to facilitate performance, more than two sources of information could be considered. This was examined by Selcon *et al.* (1995), who investigated whether redundancy gains increased when more than two sources were available. They used the example of Missile

Approach Warnings (MAW) where avoiding the missile takes short-term precedence over all other considerations. Their paradigm used an exhaustive combination of one, two, three or four information sources to find out whether the benefits of increasing redundancy continued to increase. The information sources were combinations of two modalities (visual/auditory) and two types of information coding (spatial/verbal). Their results showed a linear increase in facilitation as the number of redundant information sources was increased, with approximately 30 per cent reductions in response times with four sources as compared to one source. The obvious implication of such a result is that, where the urgency of the warning justifies it, as many layers of redundancy as possible should be used.

The results of the Selcon *et al.* (1995) study were validated by Jackson and Selcon (1998). They tested whether a PDP model of information integration would have predicted the experimental data. Using a computer-based neural network simulation, they produced predicted data for the experimental conditions and correlated these with the experimental results. The modelled data were found to be able to account for up to 96 per cent of the variance in the empirical data, strongly suggesting that the facilitation was occurring as a result of information integration at common nodes on a neural pathway. Although such a PDP approach requires extensive validation, the Jackson and Selcon results do suggest it may have possible benefits in future warning design. The most efficient approach to design is to have a priori design tools able to predict performance and avoiding the need for continual empirical assessment of new design alternatives. The implication of being able to model accurately how different information sources would combine is that it may be possible in future to predict the most appropriate redundant combinations for a particular warning, based on its criticality within the broader mission context.

Conclusions

It can be seen from the above discussion that the concept of redundancy gains has potential benefits in the design of time-critical warnings. Extensive coverage of the relevant literature has been provided to ensure that such a conclusion is based on a firm theoretical position. Specific examples from aviation warnings are also given to show how the predictions of the theory can be carried through to provide solutions to concrete, real-world warning contexts. The ability to take the concept further into a predictive design tool has also been considered. Thus it seems clear that the use of redundancy should become an integral part of the design of future warning systems.

References

Carswell, C.M. and Wickens, C.D. (1990) 'The perceptual interaction of graphical attributes: Configurality, stimulus homogeneity and object integration', *Perception and Psychophysics*, **47**, 157–68.

Chmiel, N. (1984) 'Phonological recoding for reading: The effect of concurrent articulation in a Stroop task', *British Journal of Psychology*, **75**, 213–20.

Cohen, J.D., Dunbar, K. and McClelland, J.L. (1990) 'On the Control of Automatic Processes: A Parallel Distributed Processing Account of the Stroop Effect', *Psychological Review*, **97**(3), 332–61.

Dalrymple-Alford, E.C and Budyar, B. (1966) 'Examination of some aspects of the Stroop colour-word test', *Perceptual and Motor Skills*, **23**, 1211–14.

Dunbar, K. and McLeod, C.M. (1984) 'A horse race of a different colour: Stroop interference patterns with transposed words', *Journal of Experimental Psychology: Human Perception and Performance*, **10**, 622–39.

Dyer, F.N. (1971) 'The duration of word meaning responses: Stroop interference for different pre-exposures of the word', *Psychonomic Science*, **25**, 229–31.

Dyer, F.N. (1973) 'Interference and facilitation for color naming with separate bilateral presentations of the word and color', *Journal of Experimental Psychology*, **99**, 314–17.

Ehri, L.C. (1976) 'Do words really interfere in naming pictures?', *Child Development*, **47**, 502–5.

Eriksen, C.W., Goettl, B., St. James, J.D. and Fournier, L.R. (1989) 'Processing redundant signals: coactivation, divided attention, or what?', *Perception and Psychophysics*, **45**, 356–70.

Farmer, E.W., Taylor, R.M. and Belyavin, A.J. (1980) 'Large color differences and the geometry of Munsell color space', *Journal of the Optical Society of America*, **70**(2), 243–5.

Garner, W.R. and Felfoldy, G.L. (1970) 'Integrality of Stimulus Dimensions in Various Types of Information Processing', *Cognitive Psychology*, **1**, 225–41.

Glaser M.O. and Glaser, W.R. (1982) 'Time course analysis of the Stroop phenomenon', *Journal of Experimental Psychology: Human Perception and Performance*, **8**, 875–94.

Glaser, W.R. and Dungelhoff, F.-J. (1984) 'The time course of picture–word interference', *Journal of Experimental Psychology: Human Perception and Performance*, **10**, 287–310.

Goettl, B.P., Kramer, A.F. and Wickens, C.D. (1986) 'Display format and the perception of numerical data', *Proceedings of the 30th Annual Meeting of the Human Factors Society*, Santa Monica, CA: Human Factors Society.

Grice, G.R. and Canham, L. (1990) 'Redundancy phenomena are affected by response requirements', *Perception and Psychophysics*, **48**, 209–13.

Grice, G.R. and Gwynne, J.W. (1987) 'Dependence of target redundancy effects on noise conditions and number of targets', *Perception and Psychophysics*, **42**, 29–36.

Grice, G.R. and Reed, J.M. (1992) 'What makes targets redundant?', *Perception and Psychophysics*, **51**, 437–42.

Grice, G.R., Canham, L. and Boroughs, J.M. (1984) 'Combination rule for redundant information in reaction time tasks with divided attention', *Perception and Psychophysics*, **35**, 451–63.

Hamers, J.F. and Lambert, W.E. (1972) 'Bilingual interdependencies in auditory perception', *Journal of Verbal Learning and Verbal Behavior*, **11**, 303–10.

Hentschel, U. (1973) 'Two new interference tests compared to the Stroop Color-Word Test', *Psychological Research Bulletin*, Lund University, **13**, 1–24. (From *Psychological Abstracts*, 1974, **51**, Abstract No. 8163.)

Jackson, M. and Selcon, S.J. (1998) 'A PDP model of information integration', *Proceedings of the First International Conference of Engineering Psychology and Cognitive Ergonomics*, Stratford-upon-Avon, 23–25 October 1996.

Jacob, R.J.K., Egeth, H.E. and Bevan, W. (1976) 'The face of a data display', *Human Factors*, **18**, 189–200.

Kahneman, D. and Chajczyk, D. (1983) 'Tests of the automaticity of reading: Dilution of Stroop effects by colour-irrelevant stimuli', *Journal of Experimental Psychology: Human Perception and Performance*, **9**, 497–509.

Kahneman, D. and Henik, A. (1981) 'Perceptual organization and attention', in M. Kubovy and J.R. Pomerantz (eds), *Perceptual Organization*, Hillsdale, NJ: Lawrence Erlbaum Associates.

Klein, G.S. (1964) 'Semantic power measured through the interference of words with color-naming', *American Journal of Psychology*, **77**, 576–88.

La Heij, W., Dirkx, J. and Kramer, P. (1990) 'Categorical interference and associative priming in picture naming', *British Journal of Psychology*, **81**, 511–25.

Logan, G.D. (1980) 'Attention and automaticity in Stroop and priming tasks: Theory and data', *Cognitive Psychology*, **12**, 523–53.

Lupker, S.J. (1979) 'The semantic nature of response-competition in the picture–word interference task', *Memory and Cognition*, **7**, 485–95.

MacDonald, J. and McGurk, H. (1978) 'Visual influences on speech perception processes', *Perception and Psychophysics*, **24**, 253–7.

MacLeod, A. and Summerfield, A.Q. (1987) 'Quantifying the benefits of vision to speech perception in noise', *British Journal of Audiology*, **21**(4), 131–41.

Marks, L.E. (1975) 'On colored-hearing synesthesia: Cross-modal translations of sensory dimensions', *Psychological Bulletin*, **82**, 303–31.

McClain, L. (1983) 'Stimulus–response compatibility affects auditory Stroop interference', *Perception and Psychophysics*, **33**, 266–70.

McGurk, H. and MacDonald, J. (1976) 'Hearing lips and seeing voices', *Nature*, **264**, 746–8.

McLeod, C.M. (1991) 'Half a century of research on the Stroop effect: An integrative review', *Psychological Bulletin*, **109**(2), 163–203.

McLeod, C.M. and Dunbar, K. (1988) 'Training and Stroop-like interference: Evidence for a continuum of automaticity', *Journal of Experimental Psychology: Learning, Memory and Cognition*, **14**, 126–35.

Melara, R.D. (1989) 'Dimensional interaction between colour and pitch', *Journal of Experimental Psychology: Human Perception and Performance*, **15**, 69–79.

Melara, R.D. and Mounts, J.R.W. (1993) 'Selective attention to Stroop dimensions: Effect of baseline discriminability, response mode and practice', *Memory and Cognition*, **25**, 627–45.

Melara, R.D. and O'Brien, T.P. (1987) 'Interaction between synesthetically corresponding dimensions', *Journal of Experimental Psychology: General*, **116**, 323–36.

Miller, J. (1982) 'Divided attention: evidence for coactivation with redundant signals', *Cognitive Psychology*, **14**, 247–79.

Mordkoff, J.T. and Miller, J. (1993) 'Redundancy gains and coactivation with two different targets: The problem of target preferences and the effects of display frequency', *Perception and Psychophysics*, **53**, 527–35.

Morton, J. (1969) 'The use of correlated stimulus information in card sorting', *Perception and Psychophysics*, **5**, 374–6.

Morton, J. and Chambers, S.M. (1973) 'Selective attention to words and colours', *Quarterly Journal of Experimental Psychology*, **25**, 387–97.

Posner, M.L. and Snyder, C.R.R. (1975) 'Attention and cognitive control', in R.L. Solso (ed.), *Information Processing and Cognition: The Loyola symposium*, Hillsdale, NJ: Lawrence Erlbaum Associates.

Raab, D. (1962) 'Statistical facilitation of simple reaction time', *Transactions of the New York Academy of Sciences*, **43**, 607–10.

Rosinski, R.R. (1977) 'Picture–word interference is semantically based', *Child Development*, **48**, 643–7.

Rosinski, R.R., Golinkoff, R.M. and Kukish, K.S. (1975) 'Automatic semantic processing in a picture–word interference task', *Child Development*, **46**, 247–53.

Rumelhardt, D.E., McClelland, J.L. and The PDP Research Group (1986) *Parallel Distributed Processing: Explorations in the Microstructure of Cognition, Vol.I: Applications*, Cambridge: Cambridge University Press.

Selcon, S.J. and Taylor, R.M. (1989) 'Evaluation of the Situational Awareness Rating Technique (SART) as a tool for aircrew systems design', in AGARD-CP-478, *Situational Awareness in Aerospace*, NATO: Neuilly-sur-Seine.

Selcon, S.J., Taylor, R.M. and McKenna, F.P. (1995) 'Integrating multiple information sources: using redundancy in the design of warnings', *Ergonomics*, **38**(11), 2362–70.

Selcon, S.J., Taylor, R.M. and Shadrake, R.A. (1992) 'Multi-modal cockpit warnings: pictures, words, or both?', *Proceedings of the Human Factors Society, 36th Annual Meeting*, vol.1, Santa Monica, CA: Human Factors Society.

Shiffrin R.M. and Schneider, W. (1977) 'Controlled and automatic human information processing: II. Perceptual learning, automatic attending and a general theory', *Psychological Review*, **84**, 127–90.

Simon, J.R. and Craft, J.L. (1970) 'Effect of an irrelevant auditory stimulus on visual choice reaction time', *Journal of Experimental Psychology*, **86**, 272–4.

Simon, J.R., Acosta, E. and Mewaldt, S.P. (1975) 'Effect of locus of warning tone on auditory choice reaction time', *Memory and Cognition*, **3**(2), 167–70.

Stroop, J.R. (1935) 'Studies of interference in serial verbal reactions', *Journal of Experimental Psychology*, **18**, 643–62.

Taylor, A. and Clive, P.B. (1983) 'Two forms of the Stroop test', *Perceptual and Motor Skills*, **57**, 879–82.

Taylor, R.M. (1984) 'Some Effects of Display Format Variables on the Perception of Aircraft Spatial Orientation', in AGARD-CP-371, *Human Factors Considerations in High Performance Aircraft*, NATO: Neuilly-sur-Seine.

Taylor, R.M. (1989) 'Situational Awareness Rating Technique (SART): The Development of a Tool for Aircrew Systems Design', in AGARD-CP-478, *Situational Awareness in Aerospace*, NATO: Neuilly-sur-Seine.

Tecce, T.S. and Happ, A. (1964) 'Effects of shock-arousal on a card-sorting test of colour–word interference', *Perceptual and Motor Skills*, **19**, 905–6.

van der Heijden, A.H.C., Schreuder, R., Maris, l. and Neerincx, M. (1984) 'Some evidence for correlated separate activation in a simple letter-detection task', *Perception and Psychophysics*, **36**, 577–85.

Zakay, D. and Glickson, J. (1985) 'Stimulus congruity and S-R compatibility as determinants of interference in a Stroop-like task', *Canadian Journal of Psychology*, **39**, 414–23.

11 Different Effects of Auditory Feedback in Man–Machine Interfaces

MATTHIAS RAUTERBERG, *IPO – Eindhoven University of Technology*

Introduction

The hearing of sounds (such as alarms) is based on the perception of events and not on the perception of sounds as such. For this reason, sounds are often described by the events they are based on. Sound is a familiar and natural medium for conveying information that we use in our daily lives, especially in the working environment (Momtahan *et al.*, 1993; Gaver, 1989). The following examples help to illustrate the important kinds of information that sound can communicate (Mountford and Gaver, 1990).

- *Information about normal, physical events* – most real-world events (for example, putting a pencil on the desk) are associated with a (soft-)sound pattern. This type of auditory feedback is parallel to the visual feedback of the actual status of all interacting objects (for example, pencil and desk).
- *Information about (hidden) events in space* – all audible signals out of our visual field (for example, footsteps warn us of the approach of another person).
- *Information about abnormal structures* – a malfunctioning engine sounds different from a healthy one and/or alarms in supervisory control environments.
- *Information about invisible structures* – all hidden structures that can be transformed into audible signals (for example, tapping on a wall is useful in finding where to hang a heavy picture).

- *Information about dynamic change* – all specific semantics of real-world dynamics (for example, as we fill a glass we can hear when the liquid has reached the top).

The textual representation of information is of most use when the operator is familiar with the domain area and can demonstrate much experience and knowledge in that area (Marmolin, 1992). In comparison, more concrete (visual and audible) representations of information that the operator can query are of most use when the area is new and unknown. Most of all, user interfaces stress visual perception. Auditory feedback can probably help to reduce eye strain. New scope for the interactive representation of complex sound-generating events and processes is possible, especially in multi-media interfaces.

The parallel use of different media and the resulting parallel distribution of information, for example by simultaneously showing a predecessor through a concrete representation and its explanation through audio distribution, lead to a denser sharing of information. In this case, the operator can dedicate his attention solely to the visual information, which has parallel audio support. This reduces the need to change the textual or other visual delivery and prevents the overflow of visual information (Edwards, 1988).

Sounds can be utilised to improve the operators' understanding of visual predecessors or can stand alone as independent sources of information. Gaver *et al.* (1991) used sounds as diagnostic support applied with the direction of a process simulation. But they did not prove the hypothesis that an interface with audible feedback is superior to an interface without sound feedback. The authors describe only some global impressions of different operator reactions to sound feedback.

In the context of supervisory control, an alarm is a signal that informs the operator of a dangerous or problematic process state. Wanner (1987) classified alarms as (a) programmed and (b) non-programmed alarms. The first alarm class is divided by Riera *et al.* (1995) into two groups: (1) breakdown alarms, that correspond to internal failures of components and (2) process alarms, that show an abnormal performance of a process. The non-programmed alarms are not defined at the time of system design, but these audible cues are used by the operator (abnormal noise, smoke, steam, explosion and so on).

Stanton *et al.* (1992) classify alarms by their input modality (visual versus auditory) and their information-processing code (verbal versus spatial). Spatial alarm processing requires a manual response to maximise performance, while a verbal alarm requires a vocal response. Typical problems with alarms are 'the avalanche of alarms during a major transient or shift in operating mode, standing alarms, alarm inflation, nuisance alarms and alarms serving as status messages' (*ibid.*, p.87).

Two experiments were carried out to estimate the effect of sound feedback. First, we investigated the effects of auditory feedback for a situation where the sound is given additionally to the visual feedback. In the second experiment we investigated the effects of auditory feedback of hidden events which were produced by a continuous process in the background.

Experiment I: Individualised Sound Feedback

We carried out a first experiment to test the hypothesis that sound feedback is particularly helpful if the user can choose his or her individually preferred sound pattern as a redundant information in addition to the visual output on the screen. To test this hypothesis we implemented a simple database system: a stack with cocktail recipes. In the sound feedback condition, the result of each database query was associated with an additional and customised sound pattern.

Method

Subjects   A total of 12 subjects (four female, eight male, mean age 22 ± 2 years) were instructed to define queries on a database to find a particular recipe. The experience of all test subjects with computer systems and cocktail shaking was equally distributed over both test conditions (determined by a test design with repeated measurements).

Material   The database has a direct manipulative interface and contains 352 different cocktail recipes. The database was implemented under HyperCard on an Apple Macintosh IIfx. The user could choose his or her preferred sound feedback with the customisation interface (see Figure 11.1). Each discrete result feedback was one of 49 different sound patterns for the following six output conditions: 'fit exactly', 'fit except one part', 'fit except 2 parts', 'fit except 3 parts', 'fit except ≥ 4 parts', 'does not fit'. The user could choose among the three classes: noise, speech and music.

Tasks   Each test subject was introduced with the following scenario: 'You are in front of your housebar. The bar contains the following components [for example, gin, rum, apricot brandy, lemon, sugar]. Which four different cocktails can be produced, each with different components?' The task was to search for appropriate cocktails when the components are given (for example, type of liquor, type of juice).

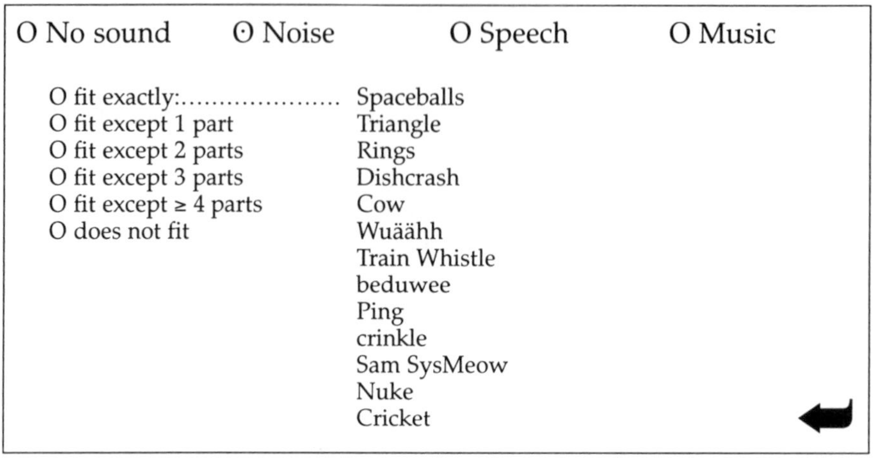

O No sound   ⊙ Noise   O Speech   O Music

    O fit exactly:...................... Spaceballs
    O fit except 1 part   Triangle
    O fit except 2 parts   Rings
    O fit except 3 parts   Dishcrash
    O fit except ≥ 4 parts   Cow
    O does not fit   Wuäähh
                          Train Whistle
                           beduwee
                           Ping
                           crinkle
                           Sam SysMeow
                           Nuke
                           Cricket   ⬅

Figure 11.1   The customisation interface to select the individual sound feedback

The database system output all possible recipes that fulfil at least one selection criterion. In general, the more selection criteria fulfilled, the better the solution was. To help the user to find the best recipe, result feedback with the customised sound pattern was provided.

Procedure   First, each user chooses individually the most convenient sound for each output condition (see Figure 11.1) from one of the three sound classes given with the sound feedback. We ran the experiment with a two-factorial test design. Factor A was 'with' or 'without' audible feedback. Test condition 1 was only visual feedback given in the form of an inverse output of all 'selection criteria' that were part of the selected recipe. Test condition 2 was visual and auditory feedback about all involved 'selection criteria'. Factor B was a repeated measurement design. Six subjects started the experiment without auditory feedback (test condition 1) and repeated a very similar task with audible feedback (test condition 2). The other six subjects started with audible feedback (test condition 2) and repeated the task without auditory feedback (test condition 1). Both task-solving trials lasted up to 15 minutes each. Each individual session took about 60 minutes.

Measures   The *first dependent variable* is the total search time (in seconds). The *second dependent variable* is the search time per recipe (in seconds). The *third dependent variable* is the number of dialog operators ('# of dialog operators'; for example, mouse clicks). Before and after

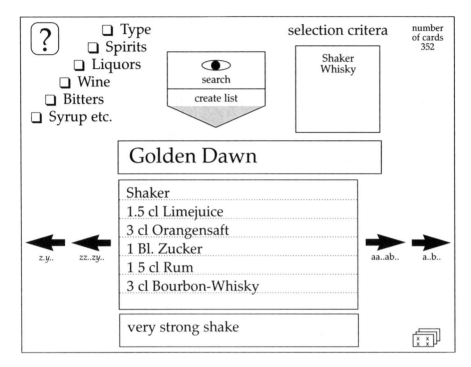

Figure 11.2   The user interface of the cocktail database

each task trial the user has to answer a mood questionnaire (eight scales with 36 items overall as monopolar rating scales). After the task trial with sound feedback we measured each personal opinion (subjective 'sound preference' questionnaire with five monopolar rating scales).

Results

The result of the personal selection, the individual customisation, shows the following distribution: speech in 42 per cent, music in 25 per cent, and noise in 33 per cent. Most of the users' sound selections did not reflect the ordinal rank structure of the search results (seven sound customisations without ordinal structure, four sound customisations with ordinal structure).

The results of the three dependent measures (total search time, search time per recipe and number of dialog operators) showed that the overall performance of direct manipulative database queries could not be significantly improved when additional, redundant feedback was given in an audible form (see Table 11.1).

In a second analysis of subjective sound preferences we distinguished between those users that prefer sound ($N = 5$) and

Table 11.1   Results of the three dependent variables that measure users' performance of database queries for both test conditions 'with sound feedback' and 'without sound feedback' (N = 12)

Variable	With sound	Without sound	P signif.
Search time total	685 ± 193	649 ± 288	0.709
Search time per recipe	201 ± 99	182 ± 113	0.627
No. of dialog operators	98 ± 38	112 ± 60	0.577

those users that do not prefer sound ($N = 7$). For this post hoc analysis we constructed the dimension 'preference' with two levels: 'sound preferred' versus 'sound refused'. Overall in both test conditions ('with' and 'without' sound feedback), when sound was preferred, the mean search time is 633 ± 248 sec; mean search time, when sound was not preferred (refused), is 692 ± 241 seconds. This difference is not significant (F $(1,10) = 0.28$, $p \leq 0.609$). But if we analyse only the test condition 'with sound feedback' then we get the following result for the variable 'search time total' between the two groups, 'preferred' versus 'refused' (see Table 11.2): mean search time for 'sound preferred' is 544 ± 215; mean search time for 'sound refused' is 786 ± 95. This difference is significant (F $(1,10) = 7.18$, $p \leq 0.023$). Similar results are found for the two other dependent variables, too.

The significant difference of the variable 'total search time' can be explained by the significant interaction term between the dimension 'test condition' and the dimension 'preference' (mean search time for 'with sound' and 'sound preferred' is 544 ± 215; mean search time for 'without sound' and 'sound prefered' is 721 ± 270; mean search time

Table 11.2   Results of the three dependent variables that measure users' performance of database queries only for the test condition 'with sound feedback' (N = 6)

Variable	Sound preferred	Sound refused	P signif.
Search time total	544 ± 215	786 ± 95	0.023
Search time per recipe	136 ± 54	246 ± 99	0.046
No. of dialog operators	71 ± 26	117 ± 32	0.029

for 'with sound' and 'sound refused' is 786 ± 95; mean search time for 'without sound' and 'sound refused' is 570 ± 310; F (1,10) = 5.04, $p \leq$ 0.049.

Discussion of Experiment I

The non-significant outcome of this comparison study could be explained by the uncontrolled factor 'individualisation': most of the users did not consider the implicit rank order in the search results. On one side we can assume a general, positive effect on the performance determined by customisation, but on the other this customisation should reflect important structures of the task itself. In the post hoc analysis we divided our test sample into two groups: users that liked their sound choices, and users that disliked their own choices. The results of Experiment I show that auditory feedback can be helpful, but only if the user chooses a sound pattern that he or she really likes.

Experiment II: Sound Feedback of Hidden Events

Our main interest in this second experiment was to test the hypothesis of Buxton (1989) and Gaver *et al.* (1991) that human operators in a 'real' process control situation monitor multiple background activities simultaneously through auditory sound feedback (*tones* as auditory and spatial alarms; see Stanton *et al.*, 1992). So we designed a system that produces audible cues and tones to help operators to monitor the status of continuous processes.

Diagnosing and treating problems with the plant were aided by alert sounds (such as breakdown alarms; see also Gaver *et al.*, 1991). We carried out this experiment to test our hypothesis in a laboratory environment with a high alarm rate during a supervisory control task.

Method

Subjects   Eight male students of computer science took part in the experiment as untrained operators (mean age of 24 ± 1 years). The experience of all test subjects with computer systems and simulation software was equally distributed over both test conditions (determined by a test design with repeated measurements).

Simulator   The simulation is based on a flexible manufacturing system that produces cases made of aluminium (see 'workpieces' in Figure 11.3). The whole system consists of eight computer numeric controlled (CNC) manufacturing centres and eight loading robots for these centres. In the input directing station all workpieces are automatically directed on the assembly line. The assembly line transports each workpiece through different stations to the CNC manufacturing centres and back to the output directing station. The whole plant was deliberately designed to be too large to fit on the computer screen, so operators could only see about half the robots and CNC machines at any time (see 'actual screen clipping' in Figure 11.3).

The status of a workpiece could be one of the following: (1) loading on the assembly line at the input directing station, (2) transportation on the assembly line, (3) fixation on the carrier at the reset station, (4) final fixation and twist on the carrier, (5) fixation on a pallet with three other workpieces at the robot, (6) processing one of two sides in the CNC station, (7) change from one side to the other at the reset station, (8) to be provided with a serial number at the labelling station, (9) loading off the assembly line at the output directing station. Steps (3) to (7) are carried out twice, once for each side of the workpiece.

We designed our simulator so that each of the machines made tones to indicate its status over time. Each tone was designed to reflect the semantics of the actual event. For instance, a splashing tone indicated that cooling liquid was being spilled. Because of the complexity of our system, as many as 38 tones could be placed at once.

We attempted to design the sounds so that none would be masked (rendered inaudible) by other sounds. Gaver *et al.* (1991) describe two strategies as being useful in avoiding masking, and we followed these. First, sounds were spread fairly evenly in frequency, so that some were high-pitched and others lower. Second, we avoided playing sounds continuously and instead played repetitive streams of sounds, thus maximising the chance for other sounds to be heard in the gaps between repetitions. CNC 0 and CNC 4 are characterised by a high-pitched sound; CNC 3 and CNC 7 are low-pitched (cf. Figure 11.3).

Normal running of a machine was coupled with a characteristic sound pattern. Each machine breakdown generated a specific alert tone, the auditory alarm, instead of the normal sound (see Table 11.3). If a robot or a CNC centre breakdown occurs, this centre cannot process the pallet of four workpieces further on. The breakdown of a machine that will not be repaired immediately leads to a jam on the assembly line. The most important – but not dangerous – consequence of an overlooked alarm is the decrease in the productivity and performance of the whole plant.

Material   We ran the experiment on an IBM compatible PC (Olivetti® i386, 25MHz, 6MByte main storage, 17" VGA colour screen) with an extra sound card (Logitech® 16Bit, 44 kHz, stereo). A special

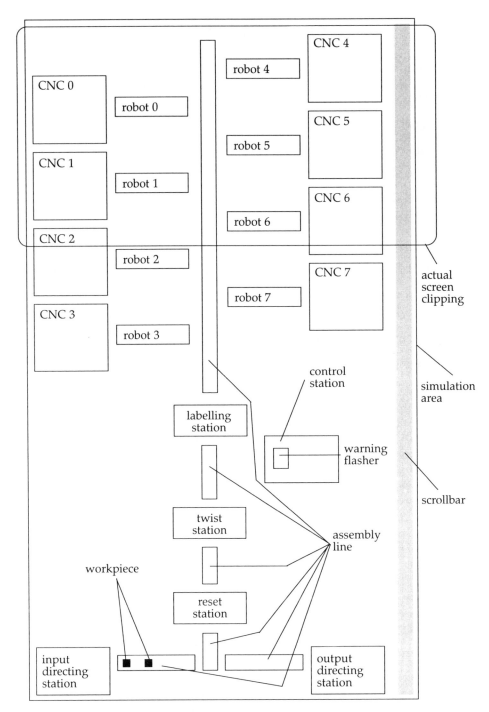

Figure 11.3 **The ground floor of the plant simulator, showing the actual screen output each operator sees at a given time**

Table 11.3 Sound types, alarm, duration and size

Machine	Sound	Alarm	Duration	Size
CNC 0-7	normal	no	1.20s	51KB
Robot 0-7	normal	no	0.39s	16KB
Input station	normal	no	0.41s	17KB
Output station	normal	no	0.78s	33KB
Reset station	normal	no	1.40s	60KB
Twist station	normal	no	0.40s	17KB
Labelling station	normal	no	0.49s	21KB
CNC 0-7	no cooling	yes	1.08s	46KB
CNC 0-7	jammed pipe	yes	1.38s	59KB
Robot 0-7	lost piece	yes	1.04s	44KB
Robot 0-7	tear off pipe	yes	1.04s	44KB
Control station	warning	yes	0.24s	10KB

Note: KB = kilo byte.

simulation program was developed in Turbo Pascal® 1.0 to present the signals on the screen. Operators heard the auditory alarms from two small active speakers (maximal 3 watt). All machines on the left side (see Figure 11.3) could be heard out of the left speaker. The right speaker gave out the sound of all machines on the right side.

Task   Subjects were instructed to operate a plant simulator and to try to maintain a high productivity rate. The task was to trouble-shoot the whole manufacturing system. First, each subject had to detect that a breakdown had happened. Then he had to find the interrupted machine (robot or CNC machine). The actual breakdown event shows the operator how to repair the machine. The operator can get this information visually in a modal dialog box with the status report at the control station or in an audible form through auditory alarm feedback. The choice of either of these possibilities was a tradeoff for the user: the cost for the visual feedback of the status report was the effort of scrolling to the control station; the cost for the auditory feedback was the memory load to discriminate the different sound pattern.

A CNC machine could have two breakdown events ('jammed outlet pipe of cooling agent', 'empty cooling agent'). A robot could break down with two different events ('lost workpiece', 'tear off a pressure pipe'). Each interrupted machine could be repaired by entering an appropriate repair code (a four-digit number, see Table 11.4) in a repair dialog box located on the machine. The operator sees only a part of the whole plant (see 'actual screen clipping' in Figure 11.3). He moves the actual screen up and down by clicking with the mouse in the scrollbar area to 'go to' the interrupted machine. A mouse click on the machine symbol pops up the repair dialog box. Entering the correct repair code returns the interrupted machine to the normal state. If an incorrect repair code is entered, no internal state change happens and the operator will hear only a short beep.

The operators' view of the plant behaviour was that robots and CNC centres break down accidentally. The plant simulator was programmed so that all breakdowns appeared in the same sequence. This approach guarantees that the trials between operators are maximally comparable.

Procedure We ran the experiment with a two-factorial test design. Factor A was 'with' or 'without' audible feedback. Test condition 1 was only visual alarm feedback with a warning flasher and a modal dialog box with status information of each manufacturing system located at the control station. Test condition 2 was visual and/or auditory alarm feedback of each machine breakdown. Factor B was a repeated measurement design. Four subjects started the experiment without auditory alarm feedback (test condition 1) and repeated the same task with audible (and visual) feedback (test condition 2). The other four subjects started with audible (and visual) feedback (test

Table 11.4 All breakdown types that lead to an alarm, and their repair codes

Machine	Breakdown	Code
CNC 0-7	no cooling	3713
CNC 0-7	jammed pipe	8319
Robot 0-7	lost piece	1731
Robot 0-7	tear off pipe	1733
Control station	status request	8700

condition 2) and repeated the task without auditory alarm feedback (test condition 1).

Each subject filled in a questionnaire to estimate individual experiences with computers (about 10 minutes). The subjects were introduced to operating the simulation tool through 'learning by using' (about 15 minutes). The simulation ran for the trouble-shooting task exactly 20 minutes. Before and after each trouble-shooting task the operator had to answer a mood questionnaire (eight scales with 36 items overall as monopolar rating scales). This mood questionnaire measures the mental workload at a rough estimate. After each trouble-shooting task we measured the subjective satisfaction with a semantic differential (11 bipolar items). Each individual session took about 90 minutes.

Measures   Our first dependent variable is a point scale that measures the productivity of the plant. Each workpiece that entered the assembly line at the input direction station counts one point. One point is counted for each side that was processed at a CNC machine (see 'Results', below). Each workpiece that left the assembly line at the output direction station counts an extra point. Each workpiece on the assembly line counts one to four points. The productivity score after 20 minutes' simulation time is the sum of all workpieces that entered the assembly line.

The second dependent variable is the number of requested status reports at the control station. The third and fourth dependent variables are number of correct and number of incorrect repairs. The eight scales of the mood questionnaire and the 11 items of the semantic differential are dependent variables to measure operators' satisfaction.

Results

First we present the results of the four dependent variables that measure operators' trouble-shooting activities. We find a significant difference between the two test-conditions for two of four dependent measures ('productivity score' and 'no. of status reports'; see Table 11.5). Without auditory alarm feedback operators moved to the control station and requested the status report significantly more than in the test condition with sound feedback (see Table 11.6). We observed that most of the operators in test condition 2 with auditory alarm went first to the control station to look for all breakdowns (visual feedback) and after that went through the whole plant to repair machine after machine. During this walk through they could identify all machines which had to be repaired by listening to the different sound pattern of each alarm type.

Table 11.5 **Results of the four dependent variables that measure operators' trouble-shooting activities for the two test conditions, 'with auditory alarm' or 'without auditory alarm'**

Variable	With alarm	Without alarm	P signif.
Productivity score	70 ± 5.6	65 ± 5.3	0.052
No. of status reports	17 ± 5.8	23 ± 4.0	0.032
No. of correct repairs	36 ± 2.5	36 ± 2.3	0.999
No. of incorrect repairs	16 ± 11.0	9 ± 7.1	0.184

On the one hand, we observed a significant improvement through auditory alarm feedback; on the other, we found that operators perceived the simulation with auditory alarms as more opaque and feel slightly more confused than without auditory alarms (see Table 11.6).

Operators felt significantly more self-assured and more socially accepted after working with auditory alarm feedback than without auditory feedback (see Table 11.7). Their readiness for endeavour, restfulness and mood improved in the test condition with sound. If we assume that the mood questionnaire measures the mental workload at a rough estimate we can suppose that in this investigation the auditory alarm feedback does not increase the mental strain.

Discussion of Experiment II

The results of this experiment give strong support for the assumption that continuous auditory feedback of hidden events can improve the productivity of man–machine interactions. Not only can performance be improved, but also some mood aspects (such as 'self-assurance' and 'self-acceptance'). One possible explanation of this outcome can be found in the way that the continuous auditory feedback helps the operator during his 'walk through the plant' to remember which machine has to be repaired next, even if the total sound pattern is complex (for example, several alarms overlapping with all other normal sounds).

Table 11.6   Results of the 11 items of the semantic differential for the two test conditions, 'with auditory alarm' or 'without auditory alarm'

Variable (-)	(+)	With alarm	Without alarm	P signif
time-consuming	time-saving	−1.1 ± 0.7	−1.0 ± 0.9	0.791
rigid	flexible	−0.9 ± 1.3	−0.8 ± 0.8	0.735
circumstantial	simple	+0.5 ± 2.3	+0.4 ± 3.1	0.889
intransparent	transparent	+0.4 ± 1.1	+1.4 ± 0.6	0.064
confused	unequivocal	+0.1 ± 2.7	+1.1 ± 1.0	0.179
unclear	clear	0.0 ± 2.6	−0.4 ± 1.4	0.596
complicated	uncomplicated	0.0 ± 1.1	−0.3 ± 1.9	0.712
prescribed	free	−0.5 ± 0.9	−0.4 ± 1.1	0.816
unforeseeable	foreseeable	0.0 ± 2.3	+0.1 ± 1.8	0.871
unsusceptible	susceptible	−0.8 ± 1.1	−0.9 ± 1.0	0.781
angry	pleased	−0.4 ± 1.7	−0.1 ± 1.3	0.709

Note: Bipolar rating scale: −2, −1, 0, +1, +2.

General Discussion

The sense of hearing is an all-round sense. This aspect is an important difference from visual perception, which is a directional sense. An auditory interface can be much larger than the visual interface (screen). Visually hidden aspects of parallel processes in the background can be made perceptible with auditory feedback (Cohen, 1993). The results of our experiment support this design approach. Auditory feedback of concurrent processes that are important for task solving improves the usability of interfaces.

Hearing is a spatial sense; we can be aware simultaneously of many sounds coming from different locations. But spatial patterns in hearing are much more limited than those of vision. It is primarily a time sense, for its main patterns are those of succession, change and rhythm. Auditory feedback typically arrives sequentially in time, whereas visual patterns may be presented either sequentially or simultaneously. Of course many perceptual experiences depend on

Table 11.7   Results of the differences (after–before) in the eight scales of the mood questionnaire for the two independent test conditions, 'with auditory alarm' or 'without auditory alarm'

Variable	With alarm	Without alarm	P signif.
Readiness for endeavour	+2.4 ± 4.1	–0.5 ± 4.1	0.199
Restfulness	+1.3 ± 2.7	+0.4 ± 3.3	0.589
Readiness for contact	+0.9 ± 2.5	–0.8 ± 2.2	0.219
Drowsiness	–1.1 ± 2.4	–1.5 ± 3.2	0.801
Self-assurance	+1.8 ± 2.0	–0.6 ± 1.7	0.022
Social acceptance	+0.1 ± 1.0	–1.1 ± 1.0	0.031
Feeling excited	0.0 ± 6.1	–1.0 ± 5.9	0.738
Mood-laden	+1.3 ± 2.2	–0.3 ± 1.0	0.128

Note: Monopolar rating scale.

the operation of several senses at once; then the prominence of one sense over another becomes a matter for study (Hartman, 1961).

Auditory feedback has poor 'referability', meaning that sounds usually cannot be kept continuously before the operator, although they can be repeated periodically. Visual patterns offer good referability, because the information can usually be 'stored' in the display. One possible advantage of auditory feedback is its 'attention-demanding' character; it 'breaks in' on the attention of the operator. Visual stimuli, on the other hand, do not necessarily win this captive audience. The operator has to be looking towards the display in order to perceive the stimulus. Finally hearing is somewhat more resistant to fatigue than sight (McCormick, 1957, p.427).

How many different concurrent tones can be discriminated? Operators reacted to up to 38 different tones in our simulation study. Momtahan *et al.* (1993) showed that staff in operating rooms were able to identify only a mean of between 10 and 15 of the 26 alarms. Nurses were able to identify only a mean between 9 and 14 of the 23 alarms found in their intensive care unit. Momtahan *et al.* put down their results to the poor design of auditory warning signals. Standardisation of auditory feedback can minimise this perceptual problem.

Cohen (1993) found that it is a difficult task to design tones 'which tell the right story and are also pleasant and emotionally neutral'. Good auditory feedback needs sound patterns that are interpretable without visual redundancy (for example, door creaks open, door slams). We have to look for sound patterns that 'stand for themselves'. Given these sounds, we have to apply them in a metaphorical sense to new events introduced by technology (for example, door creaks open => login, door slams => logout; see Cohen, 1993). For simulation tools that deal with real world events, we can easily use the corresponding real-world sounds.

The results of our study support the 'real sound' approach. To avoid boredom and fatigue – caused by outputting always the same sound pattern – the design of tones for auditory feedback should be highly context-sensitive. For example, listening to everyday sounds is based upon the perception of events and not upon the perception of sounds *per se*. This fact becomes clear in the following example (Rauterberg *et al.*, 1994): 'A pen dropped upon a piece of paper from a height of about 15cm creates a different sound than when it is dropped upon the hard surface of a desk. An altogether different sound is created when a rubber eraser is dropped upon the paper or, respectively, on the desk.' The sound created in each case is neither a characteristic of any of the participating objects (pen, rubber eraser, sheet of paper, desk surface) nor a characteristic of the event 'dropped' itself. The four different sounds in the examples are, with an observation that holds true to the reality of the situation, solely determined by their respective interaction and environmental conditions. Most of the natural sounds are a result of one or more interactions between two or more objects in a definite place and in definite surroundings and can be defined as follows:

$$\text{Auditory feedback} = f \text{ (process objects, interaction, process environment)}$$

Every interaction possesses attributes that have an influence on the produced sound (cf. Darvishi *et al.*, 1995). A framework concept for the description of auditory feedback is needed in which auditory alarms can be represented as auditory signal patterns along several descriptive dimensions of various objects interacting together in a certain environment (cf. Munteanu *et al.*, 1995). This approach is appropriate especially for the design of auditory feedback signals of the process alarms. To make auditory alarms context-sensitive leads directly to a design strategy that reduces the number of context-free alarms (cf. the discussion of 'reduction techniques' in Stanton *et al.*, 1992).

Conclusion

The results of our experiments showed that the performance of operating a plant simulator could be significantly improved when feedback of machine breakdowns and other disturbances was given in an audible form, too. We can also observe a significant enhancement of different aspects of operators' mood. Overall we can say that operators feel better and less stressed with sound feedback than without.

We found that auditory alarm feedback was effective in the following way. It helped operators keep track of the processes that were going on. Auditory alarms allowed operators to track the activity, rate and functioning of normally running machines. Without auditory feedback, operators overlooked machines that were broken down. With auditory feedback, these problems were indicated either by the machines' sound ceasing or by the various alert sounds. Continuous auditory feedback allowed operators to hear the plant as an integrated complex process. The sounds merged to produce an auditory pattern, much like the many sounds of everyday machines.

Using non-speech sounds to provide system information is appealing, for several reasons. First, by adding sound to the interface the bandwidth of communication can be significantly increased. Second, the information conveyed by sounds is complementary to that available visually, and thus sound can provide a means of displaying information that is difficult to visualise, especially with limited screen real estate. Auditory alarm feedback can also help to improve the usability of interfaces: most interfaces stress visual perception, so that auditory feedback can help to reduce eye strain and fatigue.

Acknowledgments

I have to thank the following people for their generous support: E. Styger, A. Baumgartner, A. Jenny and M. de Lange for developing the software, and all students participating as test subjects.

References

Buxton, W. (1989) 'Introduction to this special issue on non-speech audio', *Human–Computer Interaction*, 4(1), 1–9.
Cohen, J. (1993) '"Kirk Here": Using Genre Sounds to Monitor Background Activity', in S. Ashlund, K. Mullet, A. Henderson, E. Hollnagel and T. White (eds) *INTER-CHI'93 Adjunct Proceedings*, New York: ACM.
Darvishi, A., Munteanu, E., Guggiana, V., Schauer, H., Motavalli, M. and Rauterberg, M.

(1995) 'Designing environmental sounds based on the results of interaction between objects in the real world', in K. Nordby, P. Helmersen, D. Gilmore and S. Arnesen (eds), *Human–Computer Interaction – Interact'95*, London: Chapman & Hall.

Edwards, A. (1988) 'The design of auditory interfaces for visually disabled users', in E. Soloway, D. Frye and S. Sheppard (eds), *CHI'88 Conference Proceedings, 'Human Factors in Computing Systems'*, New York: ACM.

Gaver, W. (1989) 'The Sonic Finder: an interface that uses auditory icons', *Human–Computer Interaction*, 4(1), 67–94.

Gaver, W., Smith, R. and O'Shea, T. (1991) 'Effective sounds in complex systems: the ARKola simulation', in S. Robertson, G. Olson and J. Olson (eds) *CHI'91 Conference Proceedings, 'Reaching through Technology'*, Reading, MA: Addison-Wesley.

Hartman, F. (1961) 'Single and multiple channel communication: a review of research and a proposed model', *Audio-Visual Communication Review*, 9(6), 235–62.

Marmolin, H. (1992) 'Multimedia from the perspective of psychology', in L. Kjelldahl (ed.), *Multimedia: systems, interaction and applications*, Berlin/Heidelberg: Springer.

McCormick, E. (1957) *Human Engineering*, New York: McGraw-Hill.

Momtahan, K., Hetu, R. and Tansley, B. (1993) 'Audibility and identification of auditory alarms in the operating room and intensive care unit', *Ergonomics*, 36(10), 1159–76.

Mountford, S. and Gaver, W. (1990) 'Talking and Listening to Computers', in B. Laurel and S. Mountford (eds), *The Art of Human–Computer Interface Design*, Reading, MA: Addison-Wesley.

Munteanu, E., Guggiana, V., Darvishi, A., Schauer, H., Rauterberg, M. and Motavalli, M. (1995) 'Physical modelling of environmental sounds', in F. Pedrielli (ed.), *Proceedings of 2nd International Conference on Acoustic and Musical Research – CIARM '95*, Ferrara: Universita di Ferrara.

Rauterberg, M., Motavalli, M., Darvishi, A. & Schauer, H. (1994) 'Automatic sound generation for spherical objects hitting straight beams based on physical models', in T. Ottmann and I. Tomek (eds), *Educational Multimedia and Hypermedia. (Proceedings ED-MEDIA'94)*, Charlottesville: Association for the Advancement of Computing in Education.

Riera, B., Vilain, B., Demeulenaere, L. and Millot, P. (1995) 'A proposal to define and to treat alarms in a supervision room', *Preprints of the 6th IFAC/IFIP/IFORS/IEA Symposium on Analysis, Design and Evaluation of Man–Machine Systems*, at Massachustts Institute of Technology, Cambridge, MA, 27–29 June 1995.

Stanton, N.A., Booth, R.T. and Stammers, R.B. (1992) 'Alarms in human supervisory control: a human factors perspective', *International Journal of Computer Intergrated Manufacturing*, 5(2), 81–93.

Wanner, J.C. (1987) 'Facteurs humains et sécurité. Séminaire: Erreurs humains et automatisation', Ecole Nationale de l'Aviation Civile, 19–20 May 1987.

12 Speech-based Alarm Displays

NEVILLE A. STANTON, *University of Southampton* and **C. BABER**, *University of Birmingham*

Introduction

Alarms are an essential part of the control interface in a wide range of domains. Their function is to inform operators when certain system parameters exceed tolerance limits, for example in industrial process control, or when changes in system state are approaching critical levels, as with the approach of enemy aircraft in military surveillance. The information in an alarm is triggered by a discrete event which is to be communicated to the operator, whereas in other displays (such as engine temperature, reactor pressure or radar detection) information is continuously available (or can be made available on demand) and is constantly being updated. In order for alarms to fulfil their role effectively, certain criteria need to be met. The 'trigger' parameters for the alarm need to be defined and tolerance limits set (Usher, 1994). More importantly, the alarm media need to be designed to communicate the information effectively to the operator. Effective communication requires a display which presents an unambiguous message in a form which can be easily and quickly comprehended by the operator. There is still much debate on the best way to achieve effective communication of alarm messages (Stanton, 1994a). Typically alarm information is communicated to the operator using a variety of media: for example, auditory alarms, annunciator panels (translucent tiles on which alarm messages are inscribed, which are illuminated when the alarm occurs), mimic displays (a graphical depiction of the process, plant or system highlighting the item affected) and text-based displays (textual presentation of the alarm message on a visual display screen).

With the increased use of computers for command and control in systems with limited space, such as reconnaissance aircraft and submarines, and with the trend towards decreasing the size of control

243

rooms in industrial process control, there is a move away from traditional hard-wired displays. This means that annunciator panels and mimic displays are being replaced by computer-based displays. In some cases, computer screens are used to display information in a format which was previously presented on a large display, as with screen-based mimic displays. In other cases, new forms of display are being employed, such as the substitution of text-based displays for annunciator panels. Given the amount of information presented on visual display units (VDUs), it might be felt that providing additional information, such as alarms, would place an unwarranted burden on operators. The process page displays on VDUs in petrochemical industries, for instance, may contain up to 800 pages of information (Stanton, 1991). Providing a page solely for alarms in such a system could easily lead to the information being missed. By way of compensation, contemporary systems may employ three VDUs for each workstation. One of the VDUs provides overview information, often including alarms, while the others are used for control operations. In addition to the 'three-VDU workstation', control rooms are characterised by the need for operators to move between workstations, either to consult other operators or to cover for colleagues. Thus systems may come about which have a very high visual information load and which have a requirement for operator mobility. This would appear to be an excellent candidate for auditory alarms. However, given the variety of alarms which can be anticipated in the control room, it is unlikely that a set of auditory tones would suffice; rather there would be a need to employ alarms based on speech. In this chapter, a comparison of text-based and speech-based alarms displays is conducted in order to evaluate the potential for speech in control room operations.

Verbal Alarm Media

Alarm media can be distinguished on the basis of three properties: channel, access and duration (Stanton *et al.*, 1992). The channel refers to the human sensory modality of the communication, such as auditory or visual. In the case of speech and text display, although alarm messages can have exactly the same content, their channel will obviously differ. The second factor, access, refers to the means by which operators can call up, or access, the alarm message; for example, the message could be presented without operator intervention or could be made available for the operator to call at a later time. It is, of course, possible to combine the two. However an important factor in access is how the operator is alerted to the fact that an alarm message is available. To a certain extent, this returns us to the discussion of channel; it might be anticipated that the auditory channel would be more alerting than the visual channel, especially in a task involving high visual load. The third factor, duration, refers to the time for which the message is presented. Scrolling-text displays are displaced by incoming messages, but remain on the screen (at least, until they reach the bottom of the list). Speech, on the other hand, is transitory. This means that, unlike text-based displays, it needs to be attended to immediately, otherwise it will be lost (unless the operator can call for the message to be repeated or the message is written down, but writing materials may not always be to hand or easily accommodated in the workspace).

Verbal alarms, whether text-based or speech-based, share several characteristics, including the following:

- they normally consist of several alarms, rather than a single message;
- the time of their occurrence can be important for diagnosis;
- they contain complex information, rather than simple messages;
- they tend to be grouped, for example in terms of time of arrival or in terms of the location of the fault;
- they tend to arrive in bursts of messages relating to a specific alarm.

Stanton (1991) points out that few alarms are genuine (as few as 1 per cent of the alarms presented in power generation and manufacturing domains were genuine, the remainder were either 'false alarms' or confirmatory messages). If speech was reserved solely for genuine alarms, it might prove effective. However the main problem for speech displays is auditory clutter, which will occur when several alarms are presented simultaneously (Baber, et al., 1992).

Text-based displays present alarm messages on a VDU page within a list of other messages. Typically the message will contain the time of occurrence of the alarm, the plant area and item affected, the nature and severity of the fault and the current status of the message (whether it has been accepted or not). Often the most recent message is presented at the top of the display and the list scrolls downwards, although this is not always the case and there is no research advising which format is most appropriate. To accept a message, the operator is required to select the relevant message and press an 'accept' key. This action changes the status of the alarm message from 'unaccepted' to 'accepted'. It is worth noting that the alarm message list can cover several pages. Thus the text-based display represents a signal to which the operator responds. The operator then seeks further information from a graphical display.

If the role of the alarm display was simply to capture the operators' attention and specify the location of a fault, it might be felt that an auditory display would have superior conspicuity to visual displays, for example, the auditory display would be omnidirectional and would break in on whatever activity the operator was performing. Thus, in workstations with high visual load and minimal physical space, auditory alarms could be extremely useful. Cowley *et al.* (1990) provide a set of 11 guidelines for incorporating speech rather than text into an application. Briefly they compare speech with text displays and suggest that, if a message is short and simple, and if it calls for immediate action without the need for remembering the message, speech should be used in preference to text. Typically alarms are supposed to elicit immediate responses from operators and are intended to draw the operators' attention to alternative information sources. Thus speech-based alarms would appear to be preferable to text-based alarms on these criteria.

Multiple Resources and Verbal Alarm Displays

It has been suggested that verbal information is an inappropriate medium for tasks involving fault diagnosis (Robinson and Eberts, 1987). This suggestion is based on the notion that operators' knowledge of the system in which they are working is primarily spatial in nature. In this respect, the display of the system is most effective if it is graphical, for example, in a mimic display. However this does not tell us how an operator's attention should be attracted to a fault. While colour coding or flashing could be used to call attention to an item on a specific page, given the number of pages with which the operator has to deal, it is important to first call the page to the screen. Automatically calling up a page with an alarm will have a number of obvious and severe consequences for operator performance. Thus there is a trend to provide text-based alarm displays, which provide the operator with the information that an alarm condition is present and with information regarding the alarm parameter in question.

An interesting question is whether it might be anticipated that a verbal alarm will interfere with the performance of a visual–spatial task, such as monitoring an industrial process. Payne *et al.* (1994) have demonstrated that spoken messages can interrupt some types of visual–spatial task but have no effect on others. Furthermore the extent of disruption appears to be related to the intelligibility of the speech used. For instance, spoken messages with an intelligibility of less than 40 per cent impaired performance on tasks involving either mathematical reasoning or spatial decision making (such as comparison of the size of shapes, to say if they are the same or different). However there was no impairment when the intelligibility level rose above 50 per cent, nor was there impairment on tasks involving tracking. The explanation of these findings is that, when intelligibility falls below 50 per cent, attending to spoken messages represents a demand on verbal attention (that is, in terms of multiple resource theory; Wickens, 1992) and this demand will impair performance on tasks with a decision-making component. This introduces the question of how synthesised speech can be expected to impair performance.

Baber, et al. (1992) have shown that performance using synthesised speech is inferior to that using human speech, especially if the tasks involve processing of the message rather than simply spotting words in the message. Thus synthesised speech can be considered analogous to low intelligibility speech, and it might be anticipated that it will have an impact on performance of decision-making tasks, such as fault diagnosis. In order to explore this issue with reference to the design and use of alarm displays, it is important to consider the nature of alarm handling.

Alarm-initiated Activities

Alarm-initiated activities (AIA) is a generic description of alarm-handling behaviour by operators. It has been developed on the basis of research in a variety of industries (Stanton *et al*, 1992; Stanton, 1994b; Stanton and Baber, 1995) and is illustrated in Figure 12.1. AIA defines seven stages through which alarm handling can progress. The initial stage (observe) involves the operator detecting the alarm. Stanton and Baber (1995) argue that, in this stage, the operator can be either active, that is searching for alarm information, or passive, that is receiving alarm information as it is displayed. At this stage one might anticipate differences between the speed with which an operator will respond to text-based or speech-based displays.

The second (accept) stage of AIA involves the operator accepting the alarm and, hence, changing its status. Following the accept stage, the operator then proceeds to the third (analyse) stage. In this stage, the operator decides on the appropriate course of action. Often accept and analyse occur simultaneously, for example, a 'nuisance alarm' occurs, the operator analyses the alarm and accepts it by turning the alarm off. If necessary, the operator will proceed to the fourth (investigate) stage, in which the source of the alarm will be sought and its cause diagnosed. From the above discussion, one might anticipate that there will be a difference in performance at this stage, when using speech- or text-based displays. Once an alarm has been diagnosed, the fifth stage is to perform some corrective action on the process in order to remove the alarm condition, and then to monitor the change in the process, before resetting the alarm.

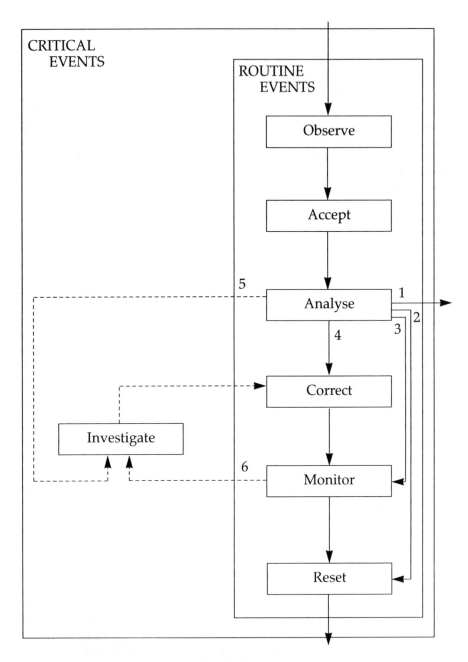

Figure 12.1   Alarm-initiated activities

This brief discussion of AIA has indicated that there may be differences between alarm media at specific stages in alarm handling.

This means that an alarm medium may be useful for one stage, but less useful for another. One potential solution to this problem would be to combine alarm media so that operators can draw on information which is presented in a fashion appropriate for a specific activity (Baber, et al., 1992). For example, Selcon *et al.* (1995) found that a combination of text and speech enhanced performance in a simulated missile warning approach task. In their study, they found a 'redundancy gain' of presenting more than one source of information, which they explained in terms of the parallel distributed processes theoretical framework.

Method

Participants

A total of 30 undergraduate psychology students (15 male and 15 female) at Aston University participated in this study. The participants were aged between 20 and 24 years. They were allocated to one of three experimental conditions (see below) such that each condition contained five men and five women.

Design

The experiment consisted of three stages: training, data gathering and a recall test. All participants performed each stage. A between-subjects design was used, with three groups each performing the same task under different experimental conditions.

Equipment

An initial training video was presented on a Ferguson Videostar videocassette recorder, using a Sharp 14" colour television. The experiment was run on an Acorn Archimedes 310 microcomputer, via a Taxan 770 14" colour monitor using mouse and keyboard for input. Software was specifically written for the experiment, and performance data were collected from the computer. Synthesised speech was generated through a synthesis-by-rule program, called 'Speech!', running on the Archimedes.

Task

The experimental task required participants to monitor a simulated industrial 'process' (see Figure 12.2) and had been used previously in Baber (1991). The process represented a simple distillery, in which

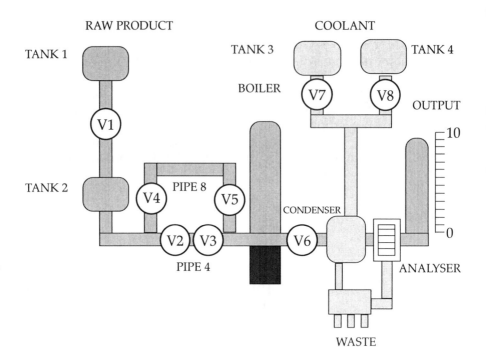

Figure 12.2   Simulated industrial process

liquid was heated to a specified temperature and then passed through a condenser to fill a tank of distilled liquid. Participants had control over the valves (which could be opened or closed) and boiler heating (which could be set to off, low, medium, high or very high). They could also inspect the status of plant elements, for example tank levels, valve positions and boiler temperature. Inspection was performed by calling up the display for a specific element. The display disappeared after three seconds on the screen. Using this information, participants were required to produce as much distilled liquid as possible. Output was shown in the final tank and the total units were measured.

There were several parts of the process which could fail: the source liquid could run out, the supply pipe could crack, the temperature of the boiler could be outside limits, the coolant liquid could run out. Each failure had an alarm associated with it. When an alarm occurred, participants were required to accept the alarm and then investigate and correct the fault. Timeliness in fault recovery was of paramount importance to overall performance on the task. As fault correction times increased, so output performance would decrease correspondingly.

In addition to the principal task of monitoring the process, participants were also presented with a spatial secondary task. Baber *et al.* (1991) had previously found this task to be affected by process monitoring. Briefly participants were presented with a figure rotation task (pairs of 'stick people' holding circles in either their left or right hands, one of the pair being rotated in either the XX, Y or Z plane relative to the other). Figures appeared in a window at the top of the screen, and participants had to indicate whether the pairs were the same or different by pressing the 'same' or 'different' keys, which were the left and right mouse buttons, respectively. The secondary task was a monitoring task, which had changing parameters and required monitoring. The intention was to add a monitoring task to the process control task in order to reflect the demands of monitoring normally associated with human supervisory control systems.

Procedure

The procedure for the experiment comprised seven steps, which can be summarised as follows:

Stage 1: training
1   Participants were presented with a 10-minute video showing a training programme describing the use of the software and the task to be performed.
2   Participants practised the task using the software in its 'unmasked' state, that is, with all plant information displayed. Practice was performed to a criterion, which was set at 80 per cent of maximum output. The training session ensured that participants could recognise the synthesised speech which, being phoneme-driven, was comparable to commercially available speech synthesis systems. The list of the alarm messages presented is indicated below:

- Boiler temperature low
- Boiler temperature high
- Tank one empty
- Tank two empty
- Tank three empty
- Tank four empty
- Pipe eight damaged
- Boiler empty
- Output tank empty
- Output tank full
- Condenser empty
- Product to waste
- Furnace tripped

In the 'speech' condition these alarm messages were presented aurally, whereas in the 'text' condition the alarm messages were presented on the screen in an alarm message box. In the 'speech and text' condition both presentation media were used.

Stage 2: experiment
3   Participants were read a set of standard instructions, explaining the nature of the task and informing them that the plant would be run in a 'masked' state. This would require participants to call up information regarding the status of plant information, using 'inspect' commands.
4   Participants monitored the process and were presented with an unanticipated, unpractised emergency. They were expected to deal with one emergency and return the process to normal operation. The spatial reasoning task was performed when the demands of the primary task allowed.
5   Participants were debriefed on the nature of the study for a period of approximately five minutes.

Stage 3: recall test
6   Participants were presented with an unanticipated recall test, in which they had to recall as many of the alarms presented during the experiment as possible.
7   Participants were thanked for their participation and given an opportunity to discuss the nature of the study should they so wish.

Measurement

Every input was logged automatically by the software. In addition, the alarms generated and the 'process output' per unit time was also logged, as was final output. Furthermore participant response times to alarm messages were recorded. The times to diagnose a fault and restore normal process conditions were also recorded. From the data recorded, it was possible to determine whether participants had engaged in any inappropriate activities or had made any keying errors during the experiment. Finally the accuracy of performance on the secondary task was recorded and participants' performance on the recall test noted.

Analysis

Performance data were analysed by one-way analysis of variance (ANOVA), with alarm media compared for output performance and times to respond to all alarms, and time to respond to and recover from the pipe break alarm as between-subject variables. The data on inappropriate activities and recall performance were analysed using a Mann-Whitney test.

Results

The results section will be divided in terms of the analyses performed. Results for the output performance are presented first, followed by the time data. Finally comparison of inappropriate activities and results for the recall test will be presented.

Output Performance

The level of output produced by participants in each of the three conditions was recorded and the mean and standard deviation of these data were calculated. For the speech alarm condition, the mean output level was 38.6 (sd = 32.5) units. For the text alarm condition, the mean output level was 74.3 (sd = 26.6) units. For the combined text and speech alarm condition, the mean output level was 64.85 (sd = 31.2) units. However there is a high level of variation in participants' performance, and an ANOVA was conducted in order to ascertain whether the apparent difference between groups was statistically significant. These data were compared using a one-way ANOVA. The results are shown in Table 12.1.

From Table 12.1 it is clear that a significant main effect ($F = 3.74$; $p < 0.05$) can be claimed for the output level data. Post hoc analysis of the data, using a Scheffé F-test, indicated that this difference was confined to the speech versus text conditions ($F = 3.486$; $p < 0.05$). Thus we can say that participants in the speech alarm condition produced significantly lower levels of output than those in the text alarm condition. However the combined text and speech alarm condition was not significantly different from the other conditions.

Table 12.1   Analysis of variance for output performance

Source	Sum squares	df	Variance	F	p
Between groups	6842.85	2	3421.43	3.744	0.0367
Within groups	24676.03	27	913.93		
Total		3518.88	29		

Alarm Handling

This section reports the results of three measures which examine participant's ability to handle alarm information presented via different media.

Time to accept alarms   The time interval between an alarm appearing and a participant manually accepting the alarm were logged by the computer. The mean time to accept an alarm in the speech alarm condition was 11.68 (sd = 12.3) seconds. The mean time in the text alarm condition was 0.29 (sd = 0.2) seconds, and the mean time in the combined text and speech alarm condition was 1.74 (sd = 2.5) seconds. Again there would appear to be a clear distinction between conditions. This proposal was tested using a one-way ANOVA. The use of ANOVA is justified by the standard deviation being proportional to the mean in all cases.

Table 12.2 shows a significant main effect (F = 7.418; $p < 0.005$). A post hoc, Scheffé F-test was used to explore this result further. Significant differences were found between speech versus text alarm conditions (F = 6.475; $p < 0.05$) and between speech versus combined speech and text alarm conditions (F = 4.933; $p < 0.05$). There was no significant difference between text versus combined speech and text conditions.

Time taken to analyse common alarms   The time taken to diagnose common alarms appeared to be quite similar across conditions, for example, the mean time with speech alarms was 7.9 (sd = 5.7) seconds, with text alarms it was 6.5 (sd = 3.6) seconds and with combined speech and text alarms it was 10.9 (sd = 6.3) seconds. A one-way ANOVA confirmed that there was no significant difference between conditions on this measure.

Time spent correcting common alarms   The mean times for this measure were also similar; for example, the mean time with speech alarms was

Table 12.2   Analysis of variance for alarm accept times

Source	Sum squares	df	Variance	F	p
Between groups	659859779.57	2	329929889.78	7.418	0.003
Within groups	1.112E9	25	44480000		
Total	1.772E9	27			

11.98 (sd = 6.8) seconds, with text alarms it was 7.74 (sd = 4.2) seconds and for combined speech and text alarms it was 12.7 (sd = 6.4) seconds. Thus one might anticipate any differences to favour the text alarm condition, but the variation in the data is sufficiently large to confound this effect. Indeed no significant difference was found between groups using a one-way ANOVA.

Fault Handling

This section reports measures relating to the participants' ability to deal with a fault presented under all conditions (a break in pipe 4).

Time to accept break   The mean time to accept the break was 8.74 (sd = 5.0) seconds for the speech alarm condition, 0.67 (sd = 0.6) for the text alarm condition and 1.4 (sd = 1.3) for the combined text and speech alarms condition. Comparison of these data was performed using a one-way ANOVA and the results were presented in Table 12.3, which shows a significant main effect ($F = 21.986$; $p < 0.0001$). A post hoc Scheffé F-test was used to compare pairs of conditions, and significant differences were found between speech versus text conditions ($F = 19.015$, $p < 0.05$) and speech versus combined speech and text conditions ($F = 15.674$, $p < 0.05$), but not between text versus combined speech and text conditions.

Fault investigation time   The mean times to investigate this fault appeared to be slightly different: for example, 4.7 (sd = 7.8) seconds for speech alarm versus 1.2 (sd = 1.5) seconds for text alarm versus 1.2 (sd = 0.6) seconds for combined speech and text alarms. The time to investigate the fault with speech alarms is over twice that of the other conditions. However one-way ANOVA indicated that the difference is not significant.

Fault correction time   The mean times spent correcting from this fault appeared to differ: for example, 10.86 (sd = 9.1) seconds for speech

Table 12.3   Analysis of variance for fault accept times

Source	Sum squares	df	Variance	F	p
Between groups	309917470.971	2	154958735.485	21.986	0.0001
Within groups	169149704.214	24	7047904.342		
Total	479067175.185	26			

alarm versus 4.4 (sd = 2.8) seconds for text alarm versus 3.6 (sd = 1.9) seconds for combined speech and text alarms. A one-way ANOVA was used to test this difference, which was statistically significant (F = 4.916, $p < 0.05$) and a post hoc Scheffé F-test did indicate a significant difference between the speech versus the combined speech and text conditions ($F = 4.236, p < 0.05$).

Inappropriate Actions

The number of inappropriate actions was compared between conditions using a Kruskal-Wallis one-way ANOVA. The mean ranks were 21 (speech condition), 14 (speech and text condition) and 10 (text condition). The resulting H value was significant ($H = 8.661; p < 0.05$) and the data were further analysed using a Mann-Whitney U-test. From this analysis, we found a significant difference between speech versus text conditions ($U = 15.5, p < 0.01$) and between speech versus combined speech and text conditions ($U = 22.5, p < 0.05$). This shows that more inappropriate actions were performed in the speech condition.

Secondary Task Performance

There did not appear to be a significant difference in performance on the secondary task, in terms of accuracy. Data were analysed using a Kruskal-Wallis one-way ANOVA and found to be non-significant. This is taken to suggest that the secondary task gave an equal loading in each condition.

Recall Test

Comparison of performance on the recall test did yield interesting differences. A Kruskal-Wallis one-way ANOVA produced an H, corrected for ties, of 11.571, which was significant: $p < 0.005$. The mean ranks were eight (speech condition), 18 (speech and text condition) and 20 (text condition). Post hoc analysis was performed using a Mann-Whitney U-test. Significant differences were found between the speech versus text conditions ($U = 7, p < 0.001$) and between speech versus combined speech and text conditions ($U = 18, p < 0.01$). This shows that participants in the speech condition were less able to recall alarm messages.

Summary of Results

The results of the preceding analyses are summarised in Table 12.4, which shows the relative significant differences between conditions on the dependent variables recorded in this study.

Table 12.4   Summary of statistical comparisons of data

Dependent variable	Condition		
	Text v. combined	Speech v. text	Speech v. combined
Overall			
Output level	ns	*	ns
Alarm handling			
Observe time	ns	*	*
Investigate time	ns	ns	ns
Correct time	ns	ns	ns
Fault handling			
Observe time	ns	*	*
Investigate time	ns	ns	ns
Correct time	ns	ns	*
Additional measures			
Inappropriate actions	ns	**	*
Secondary task	ns	ns	ns
Recall test	ns	***	**

Note: ns = No significant difference found between conditions; * = significant difference at the $p < 0.05$ level; ** = significant difference at the $p < 0.01$ level; *** = significant difference at the $p < 0.001$ level.

Discussion

The study reported in this chapter compares two verbal alarm media. The media differ in terms of channel and duration, but access and content were similar across conditions. The study demonstrates a significant effect of channel on performance. Possible explanations for this finding are the quality of the synthesised speech, the intrusiveness of the speech on the control task, or participants waiting for the speech to finish before taking action. We believe that the quality of the synthesised speech was not the cause of the performance difference, for two reasons. In a study comparing human with synthesised speech, Baber, et al. (1992) reported that there were no differences between the two conditions for a matching task. The training session enabled participants to learn to discriminate between the messages and identify the variable to which they referred. However it is clear that performance using text-based and speech-based displays differs largely at the 'accept' stage of AIA. One might feel that this was simply a function of message length, with the synthesised speech display taking an appreciable time. In other words, the results could have been confounded by differences in duration. Indeed, Ito *et al.* (1989) have shown that people often wait until a synthesised display is completed before proceeding with their task. Certainly this effect was observed during this study. However, given the nature of the alarm messages, the duration of the synthesised speech message did not exceed three seconds in length. Thus, while one would anticipate possible time effects, there is somewhere in the order of a tenfold difference between speech and text displays. This implies that the difference is due to more than simple duration.

The fact that there was a significant difference in recall when speech was used implies that the synthesised speech was of sufficiently low intelligibility to cause participants problems (see also Baber, et al., 1992). However the problem arising from intelligibility appeared to be resolved within the acceptance stage and did not have an impact on subsequent performance. From the work of Payne *et al.* (1994) one might anticipate that the speech display would lead to inferior performance on tasks involving decision making. We found significant effects in terms of both reaction time and activity: specifically the speech condition made significantly more inappropriate actions than the other groups.

The differences in performance could be explained, in part, by an appeal to multiple resource theory. For instance, the intelligibility of the synthesised speech represented an attentional demand which had an impact on task performance, in terms both of time and of activity. However the fact that there was little effect, in terms of performance time, on stages other than 'accept' suggests that participants were not using the information in the alarm message to perform subsequent tasks. However the notion of attentional demand does not allow extension to consider whether a similar effect would occur with speech of higher intelligibility. Certainly the work of Payne *et al.* (1994) would imply that simply making the speech more intelligible would increase performance. However we are not sure that this would necessarily be true for this experiment.

Given the fact that participants were able to understand the speech-based display (and that it was of similar quality to other commercially available synthesised speech), one could argue that the speech had sufficient functional intelligibility for the task. Given also the problems which will arise if the speech-based display becomes too similar to human speech (Baber, et al., 1992), one might anticipate that simply improving the quality of the speech may not be possible.

There is some evidence to suggest that irrelevant speech may have adverse effects upon performance (Salamé and Baddeley, 1982; Smith,

1989). The effects appear to occur independently of intensity, within the range of 55–95 dB(A). Consider the participant working to maintain the process while speech alarms are being presented. The alarm information may be described as 'irrelevant' if it does not relate to the particular task in hand, and performance is disrupted. Task relevancy is an important concept. While alarm information may be important for the overall task of maintaining plant production and efficiency, it may not be immediately relevant to the subtask being performed at the specific time that the alarm is presented. Stanton and Baber (1995) identify two types of alarm observation activities: active extraction and passive reception. Active extraction is characterised by operators sampling displays, searching out information that relates to their current goals and intentions. Passive reception is characterised by the alarm attracting the attention of operators and forcing them to focus on changes in the system's status. Arguably the participants have no control over the sampling of speech alarm messages and this can force them into the passive reception mode. The lack of control over sampling and task irrelevancy of information presented may lead to increased attention and memory demands. These increased demands may lead to a general performance decrement. The differences in recall performance indicate that participants in the 'speech' alarms condition did not commit the messages to memory to the same degree as the participants in the 'text' alarms condition.

The main difference in performance occurs at the 'accept' stage, resulting from a delay between the alarm and action in the speech condition. This delay can be said to parallel a phenomenon termed 'cognitive lock-up' (Moray and Rotenberg, 1989). Operators may observe abnormal values, but fail to act because they are already busy dealing with another fault and may wish to finish dealing with one problem before tackling a new one. If the speech display places an attentional demand on the operator, dealing with the alarm display itself becomes a problem. This would result in not only a time delay pertaining to the length of the speech message, but also an additional time delay resulting from processing and handling the message.

Conclusions

It is suggested that speech alone as a medium for alarm displays cannot be recommended for tasks where there is a memory component, there is likely to be some delay before the fault is attended to, there is likely to be more than one alarm presented at a time, and the operator is required to assimilate information from a variety of sources using spatial reference. If speech is to be incorporated into the alarm system for 'process control' tasks, it is recommended that it be paired with other media such as a scrolling text display. However speech-based alarms might be appropriate for tasks where an immediate response is required, the 'operator' is away from the interface, the situation is typically one-alarm to one-event, and fault management is serial in nature. Further investigation is needed before this latter proposal can be firmly recommended.

Acknowledgments

The authors are grateful to Sally Taylor-Adams who collected the data for this study and Ray Taylor who developed the simulated process plant. The authors also gratefully acknowledge the publishers of *Ergonomics* for allowing them to reproduce the text of this chapter.

References

Baber, C. (1991) *Speech Technology in Control Room Systems: A human factors perspective*, Chichester: Ellis Horwood.

Baber, C., Stammers, R.B. and Taylor, R.G. (1991) 'An experimental assessment of ASR in high cognitive workload situations in control room systems', in Y. Quéinnec and F. Daniellou (eds), *Ergonomics for Everyone: Proceedings of the 11th Congress of the International Ergonomics Association, Paris 1991*, London: Taylor & Francis.

Baber, C., Stanton, N.A. and Stockley, A. (1992) 'Can speech be used for alarm displays in "process control" type tasks?', *Behaviour and Information Technology*, **11**, 216–26.

Baber, C., Usher, D.M., Stammers, R.B. and Taylor, R.G. (1992) 'Feedback requirements for automatic speech recognition in the process control room', *International Journal of Man–Machine Studies*, **17**, 703–19.

Cowley, C.K., Miles, D. and Jones, D.M. (1990) 'The incorporation of synthetic speech into the human–computer interface', in E.J. Lovesey (ed.), *Contemporary Ergonomics*, London: Taylor & Francis.

Ito, N., Inome, S., Onkura, M. and Masada, W. (1989) 'The effect of voice messages on the interactive computer system', in F. Klix, N.A. Streitz, Y. Waern and H. Wankle (eds), *Macinter II*, Amsterdam: North-Holland.

Moray, N. and Rotenberg, I. (1989) 'Fault management in process control: eye movements and action', *Ergonomics*, **32**, 1319–42.

Payne, D.G., Peters, L.J., Birkmire, D.P., Bonto, M.A., Anastasi, J.S. and Wenger, M.J. (1994) 'Effects of speech intelligibility level on concurrent visual task performance', *Human Factors*, **36** , 441–75.

Robinson, C.R. and Eberts, R.E. (1987) 'Comparison of speech and pictorial displays in a cockpit environment', *Human Factors*, **29**, 31–44.

Salamé, P. and Baddeley, A.D. (1982) 'Disruption of short-term memory by unattended speech: implications for the structure of working memory', *Journal of Verbal Learning and Verbal Behaviour*, **21**, 150–64.

Selcon, S.J., Taylor, R.M. and McKenna, F.P. (1995) 'Integrating multiple information sources: using redundancy in the design of warnings', *Ergonomics*, **38**, 2362–70.

Smith, A. (1989) 'A review of the effects of noise on human performance', *Scandinavian Journal of Psychology*, **30**, 185–206.

Stanton, N.A. (1991) 'Alarm information in fault diagnosis', IEE Digest number 1991/156, IEE, London, 9/1-9/5.

Stanton, N.A. (1994a) 'Key Topics in Alarm Design', in N.A. Stanton (ed.), *Human Factors in Alarm Design*, London: Taylor & Francis.

Stanton, N.A. (1994b) 'Alarm-initiated Activities', in N.A. Stanton (ed.), *Human Factors in Alarm Design*, London: Taylor & Francis.

Stanton, N.A. and Baber, C. (1995) 'Alarm-initiated activities: an analysis of alarm handling by operators using text-based alarm systems in supervisory control systems', *Ergonomics*, **38**, 2414–31.

Stanton, N.A., Booth, R.T. and Stammers, R.B. (1992) 'Alarms in human supervisory control: a human factors perspective', *International Journal of Computer Integrated Manufacturing*, **5**, 81–93.

Usher, D.M. (1994) 'The Alarm Matrix', in N.A. Stanton (ed.) *Human Factors in Alarm Design*, London: Taylor & Francis.

Wickens, C.D. (1992) *Engineering Psychology and Human Performance*, New York: HarperCollins.

PART V
PRACTICAL ISSUES

13 Designing Aircraft Warning Systems: A Case Study

J.M. NOYES, *University of Bristol, Department of Experimental Psychology*, A.F. CRESSWELL STARR, *Smiths Industries Aerospace, Cheltenham* and J.A. RANKIN, *British Airways*

Introduction

Warnings are commonly found in safety-critical systems where it is necessary for the human operator to monitor a set of parameters in order to ensure that the systems under observation are performing within defined safety limits. Often, the system being monitored is complex with many components, and subsequently there is a need to measure the behaviour of a large number of parameters. When these parameters deviate outside of predetermined fixed thresholds, the warning system directs the operator's attention to the state of the system, usually via a number of visual and/or auditory alarms. Paradoxically, most of the operator's job may be uneventful to the point of boredom – a state of affairs which changes very rapidly when an event triggers an alarm, or a number of alarms. The monitoring role assumed by the human operator then changes to that of a diagnostician; he or she must sort out and remedy what may be a complicated, multi-causal, and often stress-inducing situation. There is also a strong likelihood that the alarms have been triggered by a system state about which the operator has no previous first-hand experience.

On the flight deck, alarms constitute part of the aircraft warning systems with auditory and/or visual alerts being the first stage in capturing the operator's attention in order to direct him or her towards an abnormal situation (or a situation moving outside normal limits). Over the decades, civil aircraft warning systems have gradually evolved from little more than a fire bell and a few lights to more sophisticated aural and visual warning and display systems capable of generating a cacophony of aural signals (for example, bells, clackers, buzzers, wailers, tones, horns, intermittent horns, chimes, intermittent chimes and synthesised voices), although recent trends suggest a reduction in auditory warnings. During the jet era alone, warnings have increased from 172 on the DC 8 to 418 on the DC 10, and 188 on the Boeing 707 to 455 on the Boeing 747 (Hawkins, 1987) and they are increasing further on the newer Boeing 747-400. This trend is likely to continue with the next generation of aircraft as systems grow in number and complexity.

In the avionics application, alarms (or alerts, as they are more commonly called) are seen as an integral part of the warning system, and like the hard-copy documentation of operating procedures for recovery when warnings occur, they tend not to be considered in isolation during design, development and operation. In the context of future systems, there are a number of implications which derive from defining warning systems in this way, since it is not possible to design parts of the warning system in isolation. For example, the design of warnings and warning systems should not be considered without reference to the overall design philosophies of the organisation, operating policies and procedures, perhaps even extending to (user) practices (Degani and Wiener, 1994; Edworthy and Adams, 1996).

Civil Aircraft Warning Systems

Historical Perspective

The safety-critical aspects of operating aircraft created the need for the crew to have easy access to information concerning the state of the aircraft systems. One of the first instruments to be included in the cockpit was the engine RPM (revolutions per minute) indicator, which helped the pilot make sure that he or she was controlling the engine within the operating limitations (Chorley, 1977). In the early days of aviation, little attention would have been given to designing systems according to user requirements, primarily because the well-established ergonomic principles known today did not exist. For example, the early altimeters rotated anti-clockwise with increasing altitude because they were based on aneroid barometers, with the barometer/weather scales simply replaced by an altitude scale. This is a counter-intuitive way to display information which is showing an increase; many pilots modified similar displays on incidence meters, which were a forerunner of attitude displays, by marking the operating limits on the dial. Movement of the pointer towards these marked limits could then be taken as a sign of a potential problem.

During the Second World War, aircraft systems rapidly became more sophisticated and their complexity increased accordingly. As a result, it became apparent that human performance was an important consideration in the design of aircraft systems, especially at the point of contact between the human operator and the system – the 'user interface'. Some standardisation began to appear in the layout of instruments across aircraft type. For example, the primary flight instruments required for 'blind' or instrument flying were located together in the forward view of the crew and arranged in what has become known as the 'basic T' shape. In the 1950s the advent of electronic servo-driven instruments allowed more parameters to be measured. In an attempt to reduce pilot workload, automated systems, such as the flight director system, were developed to assist flight control and navigation (Chorley, 1976). These developments in instrumentation resulted in an increased number of warning lights associated with the aircraft systems. At this stage, the system warning lights were positioned on the appropriate systems' panels, which were located across the flight deck (Gorden-Johnson, 1991). Since warning indications were predominately visual, the crew had to scan the panels continually to check for the appearance of a warning.

Over the intervening decades, the 'master/central warning and caution' concept was developed; the approach here involved grouping together attention-getters and warning lights on a single panel within the pilots' forward visual field (Alder, 1991). These were categorised into red for 'urgent' (action needed now) and amber for 'less important' (attention needed now). The warning indicators contained a caption identifying the system with which the warning was concerned. Complete failure diagnosis required reference to the appropriate system's panel (or control) indicated by the warning. As the number of warnings increased, a range of audio tones was introduced as an aid to identifying the most significant failures. Despite this, it was still possible for the crew to become overloaded with information in the event of some complex failures. Part of the response to this problem was the introduction of the so-called 'integrated displays' in the form of electronic CRT (cathode ray tube), software-controlled displays in the 1980s.

Although sensing and data handling capabilities have been expanding more rapidly over the last decade, the number of displays fitted in civil aircraft such as the Boeing 757 and 767 and the Airbus 310 and 320 has actually declined (Starr and Noyes, 1992). This new display technology allows the presentation of detailed system information in a centralised position, using combinations of text messages, schematic displays of systems and electronic check-lists.

One of the primary features of these displays is their ability to have formats and/or elements that can change in flight to reflect the current situation and information needs of the crew. However, recent trends towards increasing automation mean that some information, particularly raw and unprocessed data, is no longer available on the flight deck.

Current Flight-deck Warning Systems

There are a number of different types of civil aircraft in current operation; these represent a range of aircraft ages, manufacturers' styles, airline cultures and operational requirements. Aircraft – and, in this context, particularly their flight decks and warning systems – have changed over time as technology has developed. With aircraft, such as the Boeing 737–100 and Concorde, having an operational life of over 25 years, there is a wide variety of aircraft types and generations within the operational aircraft fleets worldwide. This enables research to be undertaken into the evolution of civil aircraft and aircraft system designs.

In order to investigate the development of warning systems over time, the features of these systems and the developmental changes made need to be understood, since a crew could be asked to make judgments, about the suitability of the features to task. This means that the warning systems currently in use needed to be placed into a taxonomy based on their features. The classification system shown in Figure 13.1 was developed for use in this study.

The warning systems which fall into the Type A Classic Group comprise fixed legends which are illuminated when the relevant failure occurs. Within this 'classic' category aircraft can be split into those in which the warning legends are located on the systems panels for which the warning is associated (type A1) and those which have some level of centralisation within their warning system (type A2). Supplementary information for the alerts in type A systems is provided in written manuals and 'Quick Reference Handbooks' that contain lists of actions for recovery, known as checklists.

Aircraft with warning systems in the type B category have six multifunction screens on which all the principal flight information – primary flight, navigation and system information – is presented. The alerts and supplementary information (such as schematics) are presented on these displays. The crew's attention is drawn to a warning or a caution by 'master warning/caution lights' situated in the forward field of view of each crew member. In the event of an aircraft malfunction, these lights will illuminate and an appropriate message will appear on one of the system screens. The layout and location of this information depends on aircraft type, and the

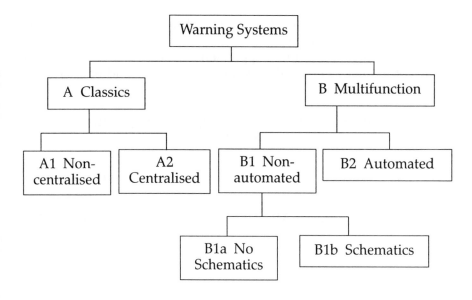

Figure 13.1 Taxonomy of warning systems in civil aircraft

provision of supplementary information relating to an alert, such as schematics and checklists, varies across aircraft types. The availability of this type of additional material on screen, as well as the means by which it 'appears', is used to break down further the classification of warning system features. Supplementary information on type B1 warning systems has to be called up by the crew, with the exception of the secondary instruments pages which appear automatically on all type B aircraft in the survey. Warning systems of type B1a have no additional supplementary information other than the secondary instruments page, whereas type B1b warning systems have system schematics that can be presented on screen. Supplementary information on warning systems in the B2 category is presented automatically when a failure condition occurs – that is, the appropriate schematic and action checklist appear on screens. These systems also have the following:

- Early Warning Triggers: when one of a number of parameters reaches a predetermined level (below the caution and warning alert level) the appropriate schematic page is presented with the trigger parameter highlighted; this acts as a pre-warning to certain conditions.
- Rule Based Prioritisation: using predetermining rules, alerts (for example, cautions) are prioritised in an attempt to highlight

primary causes. In all other multifunction warning systems, warnings are organised temporally.

● Consequences: some supplementary information about the consequences of a failure is presented at the end of a checklist of actions.

The classification in terms of the warning systems' features discussed can be summarised as shown by Table 13.1.

Table 13.1   Warning system classification by function

Type		Class	MC	C	M	CU	A	S	CH	P
Classic		A								
		A1	X							
		A2	X	X						
Multifunction		B								
		B1a	X		X	X				
		B1b	X		X	X		X		
		B2	X	X	X		X	X	X	X

Key: MC = Master Caution Warning   C   = Centralised Warnings
       M   = Multifunction Warning Displays CU = Crew Call Up
       A   = Automatic Presentation   S   = Schematics
       CH = Checklists on screen   P   = Limited within Category
                                                  Prioritisation

Looking to the Future

Despite the above advances, there are still areas in which a number of improvements might be made on current systems. This is particularly important in the light of an ever-increasing volume of air traffic – growth rates of 5 per cent (Hart, 1994) and 7 per cent (Marsh, 1994) per annum are predicted. If the volume of air traffic continues to increase annually, it is anticipated that this will be accompanied by an associated rise in the number of aircraft accidents and incidents. One suggested figure is that, by the year 2015, there will be 10 major accidents per annum (Last, 1995).

Present trends are towards the availability of an ever-increasing amount of information on the flight deck. This has implications in terms of the maintenance of crew situational awareness – a situation exacerbated by the recent tendency towards increasing automation and layering of information in multi-functional pages. The increase in automated systems and, on some of the newer fleets, the reduction in crew complement from three to two with the loss of the flight engineer are generally thought to influence the ease with which full situation awareness can be maintained. This decrease in situational awareness has been described as 'peripheralisation' by Satchell (1993) and others, and arises as a primary consequence of diminishing direct access to information about the current status and activity of the aircraft systems. However, as indicated by Pew (1994), situation awareness extends beyond merely being aware of system states to include anticipation of situations as well as interpretation of information.

Situation awareness is intrinsically linked with automation in that increased automation of systems generally results in a reduction in the extent to which crew interact with the aircraft systems. Furthermore, it also results in fewer opportunities for the crew to discuss aspects of the flight operations with each other, further exacerbating the lack of situation awareness. This aspect of automation is of increasing concern in aviation (James *et al.*, 1991) and, taken to the extreme, automation could eventually result in systems which operators will not understand (Perrow, 1984). Maintenance of situation awareness is therefore an important consideration in the design of the next generation of aircraft warning systems (Daly *et al.*, 1994).

Besides the generic issue of the maintenance of situation awareness, there are also a number of specific aspects relating to the design of warning systems *per se*. Two of these are handling multiple warnings and false warnings. Situations in which multiple and false warnings occur have been identified by recipients as being potentially difficult to manage. In terms of the latter, the implications are fairly obvious: false warnings degrade the integrity of the system and may create complacency in thinking that warnings could be false when in fact they are not. Furthermore, responding to a false warning when unsure whether or not it is indicative of an actual situation may increase the overall risk of something going wrong.

In the case of a multiple warnings situation, difficulties arise because, first, the primary failure could be masked by other failures which direct the recipient's attention (initially at least) to a non-source failure. This happens in the case of a 'cascade failure', in which a single failure causes other systems to fail, thus setting off a number of alarms and warnings. In addition, humans when problem solving tend to assume that a multiple warnings situation will arise from a common cause (Tversky and Kahneman, 1974). This may in fact be erroneous and will hinder a correct response towards recovering the situation.

The authors have recently completed a research programme which ascertained crew response to some of these issues both in current civil aircraft warning systems and a hypothetical future aircraft. Results relating to the aforementioned design aspects, as well as the methodology employed to locate this information are reported below.

Case Study: Development of a Warning System

Technological developments, combined with the continuing need for manufacturers and airlines to maximise safety and operational efficiency, were responsible for the consideration of the development of an advanced warning and diagnostic system (Noyes *et al.*, 1995). With the advent of more intelligent systems and associated programming techniques capable of collecting and utilising more information from aircraft systems, this project was precipitated by the need to ensure that user decision making on the future flight-deck would be fully supported. The focus of the study reported here revolved around the collection of information from the civil aircraft crews about their current warning systems, the rationale being that this would allow the strengths and weaknesses of current systems to be identified – information which could be used to work towards the improved design of future systems interfaces. This is discussed in detail in Noyes *et al.* (1996). It should be noted that a multi-disciplinary team worked on this research programme, including engineers, software designers, human factors psychologists and flight deck crew.

Methodology

Since the elicitation of aircraft crew needs and preferences was deemed central to understanding the good and bad features of current warning systems, the methodology employed was of critical importance. A phased methodology was chosen; this allowed access to the opinions of a large sample of crew, but at the same time did not compromise on the quality of the information collected owing to some of the specific characteristics of this user group. Many of the suggested techniques for knowledge elicitation are standard psychological tools: they can be broadly divided into 'direct methods', where experts are asked to report directly their experiences – for example, interviews, questionnaires, verbal protocols – and 'indirect methods' – such as traditional experiments, simulations, observational studies – that capture and analyse patterns of responses. In this work, a five-phase approach, as outlined by Noyes and Starr (1995), was adopted, using a variety of direct and indirect procedures. This included the following.

Phase 1: familiarisation activities   In order to encourage 'expert users' to respond to later direct elicitation activities, the experimenters had to have (and demonstrate) a sound knowledge and understanding of the expert's domain. In order to achieve this common framework, a number of familiarisation activities, including simulator visits, short- and long-haul flights, library-based research, field trips to relevant branches of NASA (National Aeronautics and Space Administration) and visits to US and European aircraft manufacturers, were undertaken by the whole research team.

Phase 2: preliminary work with crews   After the initial introduction to the users' domain with a number of flight deck crew, some introductory training on relevant aircraft systems and operation, and simulator and observation flights were undertaken during which informal and, later, more focused discussions with crews took place. These allowed identification of those design areas to be developed and studied during subsequent phases.

Phase 3: semi-structured interviews   With a view to the future development of a questionnaire, it was decided to extend the informal and focused discussions with crews, and formalise this approach into an interview schedule. One of the reasons for selecting the interview technique was to check the robustness of the questionnaire before its circulation to a wider population. Unlike questionnaires, interviews allow a two-way interchange of ideas, which was thought to be an appropriate and valuable exercise at this stage in the process. The content of the interviews was based on validating the hypothesised advantages of the proposed system, as well as a number of factors identified as important both during the work with pilots during the familiarisation phase and from analyses of flight deck documentation, such as QRHs, operations manuals, and so on.

A total of 57 questions formed the basis of in-depth, semi-structured interviews with 19 senior management and flight technical pilots, and training officers drawn from seven different aircraft types. Interviews lasted between 45 and 90 minutes and were taped with the permission of the interviewees. The data collected were not formally analysed in detail; what was of primary interest here was the detail pertaining to the questions, such as the validity of the questions, problems relating to comprehension, semantics and style, and whether it was appropriate for distribution across aircraft types.

Phase 4: questionnaire survey   Work carried out in Phases 1 to 3 facilitated the development of the questionnaire, the final format of which consisted of a biographical section plus 47 questions in the form of a statement to which respondents replied by stating the level of their agreement on a 7-point Likert scale (from '1 = Strongly Agree' through '4 = Neutral' to '7 = Strongly Disagree'), and a number of free response comment boxes.

The use of a questionnaire survey for obtaining the required data had been chosen primarily to draw on the opinions of a large number of users. Experts are known to have widely differing views (Wyatt and Emerson, 1990) – a situation which may be exacerbated by questioning only a small group of users. Consequently, the questionnaire was mailed in company packages to 3364 UK-based commercial airline pilots and flight engineers with 1359 (40.4 per cent) being completed and returned – a high response rate for a postal-type survey. Across the sample, respondents represented 10 different aircraft types: a further 'Others' category contained Hawker Siddeley 748 aircraft (n = 6), but this formed a group too small to warrant analysis, and these returned questionnaires were therefore excluded from the analyses.

Phase 5: Verification exercise   The final phase of this work with the users involved returning to a small subsample of crew in order to check the validity of the interpretation of the findings from the questionnaire data analyses. This took place both formally, in a seminar setting, and informally with the project-allocated pilots. This verification exercise was perceived as being critical in helping ensure the quality of the elicited knowledge in terms of the system application, and the ultimate goals of the research.

Results

As already mentioned, the Warning Systems Questionnaire comprised 47 questions. Several of these questions were presented in pairs so that the first question ascertained whether the respondent had this particular function on their aircraft type (the descriptive response) followed by asking respondents whether they would like this function (the prescriptive response). It is not feasible here to consider all of the questionnaire results, instead, some key questions (and salient results), relating specifically to user requirements in terms of some of the design issues raised above, will be discussed. These included the following.

Completeness of Warning Information

Statements
All warnings appropriate to the situation are given.

The warnings provided in my aircraft are usually sufficient to identify immediately the source of the problem.

Responses were generally favourable to these statements. The analysis grouped responses into two main clusters, which can be interpreted in terms of the centralisation of the warning system being used by the respondents. One group included crews flying type A aircraft, who expressed a significantly less positive response towards this feature. The centralisation on warning systems in one of the type A subgroups – A1 – requires the crew to scan the various systems panels in order to establish the presence of any master warning condition. The other subgroup – A2 – has some centralisation of warnings, but this only indicates in which system the malfunction is located. To establish the exact cause of an abnormality, crews using these warning systems have to seek further information from the relevant system panel. It is this lack of centrally located information that seems to explain why the warnings provided by all type A warning systems are perceived to be less effective at immediately indicating the source of a problem than those on which the majority of warnings are presented in a central location. The remaining respondents formed a largely heterogenous group whose warnings were centralised to some degree. In summary it could be suggested that meaningful centralisation is positively viewed as a design feature.

Statement
Warnings that are given are effective in directing me to appropriate procedures for dealing with the problem.

Responses to this statement were generally favourable: over 80 per cent of crew agreed that the warnings are effective in directing them towards appropriate actions. A significant variation was found with type B (with multifunction screen-based) warning systems, with them being regarded more highly in this respect. This may be explained by the increasing centralisation of these comprehensive warning information systems: some of them generate schematic 'pictures' of the systems, while the most advanced generate screen-based checklists of actions.

Statements
Flight deck displays provide adequate information about secondary consequences of malfunctions (e.g. inoperative systems, restrictions on operational procedures etc).

Flight deck displays should provide information about secondary consequences of malfunctions.

In this study, the only warning system type with any provision for the supply of this type of information within the warning system itself is category B2 and, as expected, the responses from the crews using this facility reflected their experience. The responses to the descriptive statement for all except crews using the type B2 system showed that they felt that their aircraft's warning system did not support the provision of information. On the other hand, responses from crews using type B2 systems were strongly positive. It should be noted that the statement refers to the flight deck display of this information and, although only type B2 aircraft have this information on their warning system displays, others may be provided with some information of this type in manuals and the like.

Responses to the prescriptive statement were uniformly positive. The mean response from crews using all warning system types was one of strong agreement with the suggestion that warning systems should provide information about the secondary consequences of malfunctions. Crews with the type B2 warning system showed the strongest agreement with this statement.

These results clearly show that crews would like secondary consequence information to be routinely provided. They also suggest that crews who have had experience with warning systems incorporating this facility feel more strongly about this issue than other crews. This could be an indication of the degree of usefulness that crews attribute to this type of information.

Multiple Warnings

Statements
It is easy to interpret warning displays when several warnings appear at the same time.

When several warning conditions are active, only the most important should be displayed.

Responses to the first of the statements on multiple warnings were neutral for the majority of crews, although a mildly negative response was made by crews flying type A aircraft. Responses from crews using type B2 warning systems gave slightly positive mean responses to the initial statement. This pattern of responses implies that there is a need to develop a system which can be easily interpreted in multiple warning situations.

Analysis of responses to the second statement on the presentation of only the most important warnings as a means of prioritising information in a multiple failure situation shows that the majority of respondents disagreed with this proposal, with the exception of those respondents using type B2 warning systems. There was a significant difference between responses from users of the type B2 systems and other users' responses, in that they were more in favour of this type of function. This may be explained by the fact that the type B2, having had experience of this feature, believe that warning systems can be built effectively to include this capability.

False Warnings

Statement
False warnings appear too often.

Evaluation of how many false warning occurrences constitutes 'too often' is highly subjective, and this subjectivity was reflected in the spread of responses to this descriptive statement, given by respondents using each of the aircraft types. As might be expected from a statement for which there was a broad spread of responses, the means were clustered around the neutral position. These results indicated that, in general, false warnings do not present a significant problem: equally, however, the results can be interpreted to imply that false warnings *do* appear too often – as one respondent replied, 'One false warning is too often'.

Anticipation

Statements
The flight-deck instrumentation available in my aircraft is effective in enabling problems to be anticipated before warnings are triggered (e.g. by indicating parameters that are slightly in error, but still within tolerance).

Flight deck instrumentation should enable problems to be anticipated.

With respect to the first statement on anticipation, there was a significant quantitative variation between the responses across the different warning system types. Inspection indicated that the responses divided into two groups – those crews with a relatively neutral attitude and those with a moderately favourable attitude. The group with the moderately favourable view flew aircraft with type B2 warning systems and those aircraft types with a flight engineer. It appears that access to detailed knowledge about systems' behaviour assists the crew in detecting changes before a warning condition occurs.

Crews from all aircraft types agreed that flight deck instrumentation should allow problems to be anticipated (from analysis of data from the second statement). By comparing each respondent's responses to the two statements, it is possible to build up a contingency table (see Table 13.2) indicating how many respondents felt that the provision of this type of information on their current aircraft is as they would like it to be.

In summary, while flight deck crews would like instrumentation to allow the anticipation of problems, not all current warning systems are perceived to provide this information.

Table 13.2   Problem anticipation

	Should be able	Should not be able
Current aircraft does allow problem anticipation	62.6%	0.5%
Current aircraft does not allow problem anticipation	33.9%	3.0%

Note: 142 respondents are not included in this table as they expressed at least one neutral attitude to the questions.

Source: Eyre *et al.* (1993).

Conclusions

The purposes of a warning system are manifold, from alerting the crew to a malfunction (or potential malfunction in the form of a discrepancy), through to providing evidence of the problem and guidelines for remedial actions. The features of a good warning system include the provision of a complete set of warnings with sufficient information (to enable anticipation of problems before they arise, and to raise awareness of them when they do arise), guidance on how to deal with the situation, provision of information about secondary consequences and the reduction of 'false warnings'.

Findings from the questionnaire study carried out with the users indicated that the majority of respondents felt that warning information was complete in terms of appropriateness of warnings to a given situation, being sufficient to identify the problem(s) and providing direction towards corrective procedures. However, in general, they considered that current warning systems provided inadequate information about secondary consequences of malfunctions, and such a provision was one which they viewed favourably. In addition, the results showed that crews would like warning systems to have a predictive capacity allowing the anticipation of problems. False warnings were generally thought to be undesirable, but were not a significant problem – presumably because they are a rare occurrence. Crews expressed a clear need for systems adept at handling multiple warnings, and it was felt that current systems were not as supportive as they could be in these situations.

Current aircraft warning systems are able to provide a large amount of data, but the presentation of this information could be improved to provide displays which better meets the needs of the crews. In general, current systems fail to:

- integrate data from several sources into a format determined by the current situation, such as phase of flight;
- allow the anticipation of malfunctions by conveying predictive information concerning abnormal conditions to the crew;
- provide advanced indication of the consequences of crew decision making and actions.

Most conventional warning systems are fault-oriented, and corrective actions are directed towards management of the immediate problem with priorities determined according to a predetermined hierarchy. Task-oriented displays may provide part of the solution in the design of future warning systems. However, prioritisation in itself is also problematic in terms of how to prioritise the information for presentation to the crew – for example, should this be done temporally as malfunctions arise, or by order of importance, implying that more critical problems should supersede those less important and so on? This has implications for the presentation of information via CRT displays.

Looking to the future, the development of warning systems on multifunction screens, utilising electronically displayed checklists, and automatically displaying system synoptics, would seem to be in keeping with users' needs as expressed in this study. One of the design strengths of the most advanced examples of this approach to warning system display technology has been identified as the direct association of warning indications and corrective actions (Alder, 1991). The capabilities of warning systems on civil aircraft have developed enormously over the past decade – a trend which will undoubtedly continue into the next century. It is important that aircraft warning systems keep abreast of technological developments, while simultaneously reflecting users' needs and requirements.

Acknowledgments

This work was carried out as part of a UK Department of Trade and Industry-funded project, IED 4/1/2200 'A Model-Based Reasoning Approach to Warning and Diagnostic Systems for Aircraft Application'. Thanks are due to British Airways for their participation in this research programme, and especially to all the flight deck crews who completed interviews and questionnaires. Thanks are also due to the late David Eyre for the meticulous statistical analyses carried out on the questionnaire data.

References

Alder, M.G. (1991) 'Warning systems for aircraft: A pilot's view', in *Proceedings of IMechE 'Philosophy of Warnings Systems' Seminar S969*, London: Institution of Mechanical Engineers.

Chorley, R.A. (1976) 'Seventy years of flight instruments and displays', *Aeronautical Journal of the Royal Aeronautical Society*, August issue, 323–42.

Chorley, R.A. (1977) *Trends in Flight Deck Displays*, Smiths Industries Report TSP-2674, 15 September, Smiths Industries Ltd., Cheltenham: Advanced Displays Studies.

Daly, K., Jeziorski, A. and Sedbon, G. (1994) 'Intelligent conversation', *Flight International*, 24–30 August, 25–27.

Degani, A. and Wiener, E.L. (1994) *On the Design of Flight-deck Procedures*, NASA Contractor Report 177642, NASA-Ames Research Center.

Edworthy, J. and Adams, A. (1996) *Warnings Design: A research prospective*, London: Taylor and Francis.

Eyre, D.A., Noyes, J.M., Starr, A.F., and Frankish, C.R. (1993) *The Aircraft Warning System Questionnaire Results: Warning information analysis*, Report 23.3, MBRAWSAC Project, Bristol: Department of Psychology, University of Bristol.

Gorden-Johnson, P. (1991) 'Aircraft warning systems: Help or hindrance philosophy of warning systems', in *Proceedings of IMechE 'Philosophy of Warnings Systems' Seminar S969*, London: Institution of Mechanical Engineers.

Hart, D.C. (1994) 'A global vision: The transition to CNS/ATM', *Avionics*, October issue, 28–31.

Hawkins, F.H. (1987) *Human Factors in Flight*, Aldershot: Ashgate.

James, M., McClumpha, A., Green, R., Wilson, P. and Belyavin, A. (1991) 'Pilot attitudes to automation', in *Proceedings of the Sixth International Symposium on Aviation Psychology*, Columbus: Ohio State University, 192–7.

Last, S. (1995) 'Hidden origins to crew-caused accidents', in *Proceedings of the IFALPA Conference, Interpilot*, June issue, 5–15.

Marsh, G. (1994) 'Towards a single-sky Europe Part 1: The present system', *Avionics*, March issue, 32–6.

Noyes, J.M., and Starr, A.F. (1995) 'Working with users in system development: Some methodological considerations', in *Proceedings of IEE Colloquium on Integrating HCI in the Life Cycle*, Digest 95/073, London: Institution of Electrical Engineers.

Noyes, J.M., Starr, A.F. and Frankish, C.R. (1996) 'User involvement in the early stages of the development of an aircraft warning system', *Behaviour and Information Technology*, 15(2), 67–75.

Noyes, J.M., Starr, A.F., Frankish, C.R. and Rankin, J.A. (1995) 'Aircraft warning systems: Application of model-based reasoning techniques', *Ergonomics*, 38(11), 2432–45.

Perrow, C. (1984). *Normal Accidents: Living with high risk technology*, New York: Basic Books.

Pew, R.W. (1994) 'Situation awareness: The buzzword of the '90s', *CSERIAC Gateway*, 5(1), 1–16.

Satchell, P. (1993) *Cockpit Monitoring and Alerting Systems*, Aldershot: Ashgate.

Starr, A.F. and Noyes, J.M. (1992) *Boeing Versus Airbus Warning System Philosophy*, Report 13.3, MBRAWSAC Project, Cheltenham: Smiths Industries Aerospace and Defence Systems, Research and Product Development.

Tversky, A. and Kahneman, D. (1974) 'Judgement under uncertainty: Heuristics and biases', *Science*, **185**, 1124–31.

Wyatt, J. and Emerson, P. (1990) 'A pragmatic approach to knowledge engineering with examples of use in a difficult domain', in D. Berry and A. Hart (eds), *Expert Systems: Human issues*, London: Chapman and Hall, 65–78.

14 The Design and Validation of Attensons for a High Workload Environment

ELIZABETH HELLIER, *City University* and
JUDY EDWORTHY, *University of Plymouth*

Introduction

This chapter reports on the design and validation work conducted to produce a set of attensons for a high workload environment. It shows how existing research was integrated into the design process and how predictions made about such designs on the basis of that research were validated through further experimentation. It also describes how laboratory experimentation can be applied to the design of auditory information in field settings and how that design process can feed back into the research base.

The warnings philosophy of the target work environment was to provide a short attenson (attention-getting sound), to communicate the priority of the event being indicated, followed by voice messages to clarify the nature of the situation. Three attensons were needed – one to indicate each of three different fault priority levels, (P1, P2 and P3). The attenson associated with each of the priority levels was required to reflect an appropriate urgency, so that the priority 1 warning was more urgent than the priority 2 warning, which in turn was more urgent than the priority 3 warning. Established principles exist whereby the perceived urgency of non-verbal auditory information can be manipulated in a logical and systematic manner (for example, Edworthy *et al.*, 1991; Hellier *et al.*, 1993). These principles were applied to the design of the attensons presented.

A constraint on the design of the attensons was that, in the interests of standardisation, the attensons designed here (the candidate attensons) should sound similar, but not identical to, attensons that were already in use in related but not identical work environments (the existing attensons). There was some scope, however, for refining the existing attensons in order to make them more urgent and possibly more discriminable from one another than was currently the case. To this end, three candidate attensons were produced for each of the three priority levels – a total of nine candidate attensons. The three candidate attensons for priority 1 were designed to resemble the existing priority 1 attenson, the three candidate attensons for priority 2 were designed to resemble the existing priority 2 attenson and the three candidate attensons for priority 3 were designed to resemble the existing priority 3 attenson. Within each priority level, previous research findings were used to design the three candidate attensons so that they differed in perceived urgency (from most to least urgent, levels a, b, and c). At each level of priority the candidate attenson most similar to the existing attenson was the least urgent version, level (c). The other versions of the candidate attensons within each priority were more urgent than the existing attenson for that priority. Besides being more urgent, the new attensons were also shorter and more discriminable than the existing attensons. In short, nine candidate attensons were designed so that they sounded similar to but not identical to the existing attensons of the corresponding priority.

Two experiments were conducted to validate the design of the attensons. In the first experiment, the perceived urgency of the nine candidate attensons was assessed. This experiment showed that, as intended, the priority 1 attensons were more urgent than the priority 2 attensons, and that the priority 2 attensons were more urgent than the priority 3 attensons. Within each level of priority it was shown that, as had been predicted, level (c) was the least urgent, followed by levels (b) and then (a). In a second experiment the nine candidate attensons were tested for similarity to the existing attensons. As predicted, it was shown that each priority of candidate attenson was most similar to the existing attenson from the corresponding priority, and that, within each level of priority, the least urgent candidate attenson (c) was most similar to the existing attenson, followed by versions (b) and then (a).

The remainder of this chapter describes the design and validation of the attensons in detail and reveals the structure of the candidate attensons. During the next phase of the project, a set of three attensons – one for each of the three priorities – will be selected from the nine candidate attensons, on the basis of trials in noise. However, some preliminary guidance can be given at this stage as to appropriate, and inappropriate, choices of combinations of attensons from the nine presented here. These recommendations are aimed at producing the appropriate gradations of urgency and discriminability between attensons at different levels of priority.

The Design Process

The Procedure

In general, the attensons were designed according to the pulse-burst principles described by Patterson (1982) and on the basis of previously conducted research into the design and perception of auditory warnings (Edworthy *et al.*, 1991; Hellier *et al.*, 1993). First, appropriate sound levels for the attensons were determined. This was achieved by establishing the masked threshold of background noise in the target work environment, and then by imposing a band 15–25dB above this masked threshold. This band gives the appropriate levels at which different components of the attensons must be placed in order to be audible without being over loud and consequently aversive. The auditory threshold at the ear was determined by four main aspects of the environment: the worst case noise spectrum from the work environment; the attenuation characteristics of the operator's headwear; the frequency response of the telephones used to transmit the attensons; and the status of the communication channel (that is, whether it was currently turned on or off).

The predicted masked threshold at the ear revealed that, in order to be heard, a low-frequency attenson component needed to be much louder (about 85dB at 200Hz) than one at a higher frequency (about 50dB at 2kHz). Two lines were drawn above the threshold curve to show the appropriate band for attenson components – 15 and 25dB above this threshold respectively. The harmonic components of each of the pulses used in the candidate attenson specifications were weighted to fit within this band. In each case, one harmonic was chosen as the standard, and all other harmonics were weighted relative to this. This standard component was given a relative amplitude weighting of 1, and all other components had values greater or less than 1, which means that their amplitude was either greater or smaller relative to the standard. Most harmonics were weighted at a value less than 1, because most of the harmonic content of the pulses was between 500Hz and 3kHz and, since masked threshold was lower here, the amplitude of the pulse did not need to be as great.

Despite the fact that low-frequency components were avoided due to the greater noise levels in this part of the spectrum, some of the attensons sounded as if they had a low fundamental frequency, because the auditory system 'fills in' in this fundamental on the basis of other, higher frequency, information. This is known as the 'phenomenon of the missing fundamental' and is an extremely useful psychological attribute under these circumstances (Licklider, 1956).

The use of the 500Hz–3kHz range for the attenson components was appropriate both acoustically and psychologically, as it allowed for five or six harmonics, which is more than enough acoustic information to ensure detection under most circumstances, provided that the attensons are set at an appropriate overall level of loudness.

The absolute level at which the attensons should be played depends on the noise spectrum of the work environment at any given time. Using a worst case noise spectrum, masked threshold at 1kHz at the ear was about 60dB. The attensons need to be presented at least 15dB above this threshold – at about 75dB at the ear. As a general guide, then, the absolute level of the attensons was recommended to be 75dB at the ear or 15dB above the masked threshold, whichever was the higher.

The second phase in the process was the specification of a small pulse of sound for each attenson, lasting from 100–200 ms. This pulse of sound typically consists of several harmonics, possibly including the fundamental frequency, to facilitate both the perception and the localisation of the attenson. The advantage of having several harmonics is that, if one or two harmonics are temporarily masked by some other sound, the pulse will still be heard because the auditory system will 'fill in' the missing parts, as previously mentioned. The possession of several harmonics also makes localisation of the attenson easier. The amplitude envelope of the pulse was also shaped so that maximum amplitude was not reached until 20 or 30 ms into the pulse, in order to avoid startle reactions. In summary, the pulse contained all the acoustic information necessary for specifying the attenson and was acoustically complex.

The next stage of construction was to specify a burst of sound for each attenson. A burst consists of several identical pulses played at potentially different frequencies and with potentially different time intervals between them. Thus the burst can be thought of as being somewhat akin to an atonal melody with a distinctive rhythm. These characteristics enable each of the final bursts to be distinctive from one another. A further stage of construction can be to create a complete warning consisting of several reiterations of the burst, with silent intervals between them, which continues until the situation signalled by the warning is attended to. For this set of candidate attensons, however, only a single burst was heard because of the relatively short time interval between the onset of the attenson and the requirement for subsequent action.

Design Considerations

Before outlining the pulse and burst parameters for the attensons, it is important to consider in more detail the design considerations affecting these specifications. In addition to calculating the appropriate level for the attenson components, as discussed above, the primary constraint on the design was the requirement for the candidate attensons to sound similar, but not identical to, existing attensons. The nature of the target work environment was such that operators moved between similar workplaces, some of which contained the existing attensons and some of which would contain the new candidate attensons. Consequently, it was important that the newly-designed candidate attensons and the existing attensons should sound recognisably the same. That is, if operators heard the existing priority 2 attenson, for example, and then at some later stage heard the candidate priority 2 attenson, they would agree that the two attensons were the same sound in principle, or at least were more similar than, say, the candidate priority 3 and the existing priority 2 attensons.

So, while there was a requirement for the candidate and the existing attensons to sound as if they belonged to the same priority 'family' or category, this did not mean that the two versions of the attensons could not – or, indeed, should not – be significantly different in several ways. Spectrally they had to be different, because the noise spectrum of the candidate attenson workplace was noticeably different to that of the existing attenson environment. Furthermore, the existing attensons were too long, and not sufficiently urgent, for the candidate attenson work environment. The candidate attensons therefore had to be shorter and more urgent than the existing attensons whilst retaining their identity as part of the appropriate priority group. Various manipulations of the attensons could be made to alter the urgency, but not the identity, of the sounds, and many of these manipulations were implemented in the design of the candidate attenson set and are discussed below.

The existing and the candidate attensons differed along two separable and important dimensions: the acoustic dimension, which includes the specification of appropriate loudness levels and the spectrum of the warning pulse; and the psychological dimension, which includes aspects of the identity and the urgency of the resultant attenson (the burst). These are dealt with in turn below.

Acoustic contrasts   As previously discussed, a warning sound will be clearly heard, without causing startle, if the warning components are somewhere between 15 and 25dB above the masked threshold of background noise for a given environment. As the noise spectrum of the candidate attenson environment differed greatly from that of the existing attenson, the appropriate band for pulse components was different for the two sets of attensons. The pulses used to make the candidate attensons contained a number of harmonics, each of which had to lie within this 15–25dB band in order to be reliably audible but not too loud. Thus some components were weighted less relative to others in order be perceived as having equal loudness. One of the main consequences of the differences in spectra between the existing and the candidate attenson environments was that different weightings of harmonics had to be specified for the two sets of attensons. Although these differences are important psychoacoustically, they were inaudible when listening to the attensons in quiet surroundings. They were, however, important when the attensons were heard in the appropriate noise environment as it was the appropriate placement of the attenson components that prevented them from being fully or partially masked by the environmental noise. The precise weightings of the harmonics are given in detail in the section on attenson specifications.

Psychological contrasts   The existing priority 1, 2 and 3 attensons have two central psychological attributes; they are clearly distinctive from one another, and they vary in their degree of urgency commensurate with their priority. This requirement is followed through in the candidate attensons and was developed further.

In order to meet the requirement for the candidate attensons to sound as if they came from the same set as the existing attensons, certain constraints were inevitably placed on the design of the candidate sounds. Despite this, each of the three candidate attensons for each priority differed from the existing attenson on which its design was based. Before they lose their identity and become unrecognisable as the same sound, attensons can vary from one another along many dimensions – their overall length, their pitch, their speed, the harmonicity of their components, their envelope shape, the number of pulses they contain, the pitch contour and, to some extent, even their rhythm. Various of these manipulations were applied to the pulses and to the bursts in the candidate attenson set in order to derive three recognisable versions of each attenson for each priority that were different from the existing attenson set (so that they could be more urgent and more discriminable) yet recognisable as members of the same category.

As an example, the existing priority 1 attenson and the most urgent candidate priority 1 attensons are compared below (Table 14.1). Although they differ along several dimensions, the candidate attenson can be identified as belonging to the priority 1 attenson family, while also being noticeably different.

In order to satisfy the requirement for the candidate attensons to be more urgent than the existing attensons, three versions of the candidate attensons were designed for each level of priority. Previous research was used to design these three versions so that they varied in perceived urgency. This was achieved primarily by manipulating

Table 14.1   Comparison of the existing attenson and one candidate attenson for priority 1

	Existing attenson	Candidate attenson
Parameter		
Pulse		
Pulse length	150ms	100ms
Envelope shape	Slow offset	Regular
Number of harmonics	7	8
Burst		
Number of pulses	6	6
Total length	900ms	600ms
Pulse frequencies	280/350Hz	280/390Hz

the speed (interpulse interval) and the pitch of the attensons, so that attensons which were required to be more urgent were faster and a higher frequency. The candidate attenson most similar to the existing attenson for each priority was the least urgency version (version (c)) and then two additional attensons were designed for increasingly urgent priority (versions (b) and (a)).

Finally, special attention was paid to the design of the priority 2 attensons. Operators demonstrated a certain degree of confusion between priority 1 and priority 2 attensons in the existing attenson set. Although this confusion was relatively small, the construction of a new set of candidate attensons presented an opportunity to eliminate it completely. Consequently, greater differences were designed between the candidate and existing priority 2 attensons than can be found for the other two priorities. However, the identity of the attenson was still preserved, as Experiment 2 shows.

The Attensons

This section gives details of each of the nine candidate attensons. For each attenson, the following pieces of information are given.

Pulses

For each pulse, the fundamental frequency (either present or apparent), the length, the onset and offset envelope, the harmonic content and the relative amplitude weightings of each of the harmonics relative to a standard harmonic (marked with an asterisk) are described. For each attenson, the same pulse is repeated to form a burst of sound, although the pitch of the pulse may be varied.

Bursts

For each burst, the number of pulse repetitions, the total duration, the pulse frequencies and the interpulse interval (milliseconds from the end of one pulse to the start of the next) are detailed. The attensons are presented in the following order: P1(c), P1(b), P1(a), P2(c), P2(b), P2(a), P3(c), P3(b), and P3(a). The details are shown in Tables 14.2 to 14.10.

Table 14.2   Attenson information: P1(c) least urgent priority 1 attenson

Pulse characteristics

Original pitch	280Hz (pitch match)	Pulse spectrum and amplitude weightings	
Harmonicity	Non-harmonic	Frequency	Amplitude weighting
Number of			
harmonics	7	1148	1*
Duration	150ms	1372	0.794
Envelope	Slow offset	1708	0.631
		1932	0.355
		2268	0.25
		2490	0.178
		2828	0.09

Burst characteristics

Number of
  pulses   6
Total length   900ms
Pulse frequencies 280Hz (2 pulses) followed by 350Hz (4 pulses)
Interpulse
  interval   0

*Standard component with amplitude weighting of 1 (used in all tables)

Table 14.3   Attenson information: P1(b) mid urgent priority 1 attenson

Pulse characteristics

Original pitch	280Hz (pitch match)	Pulse spectrum and amplitude weightings	
Harmonicity	Non-harmonic	Frequency	Amplitude weighting
Number of			
harmonics	8	1092	0.79
Duration	100ms	1148	1*
Envelope	Regular	1372	0.794
		1708	0.631
		1932	0.355
		2268	0.25
		2490	0.178
		2828	0.09

Burst characteristics

Number of	
pulses	6
Total length	600ms
Pulse frequencies	280Hz (2 pulses) followed by 350Hz (4 pulses)
Interpulse	
interval	0

Table 14.4   Attenson information: P1(a) most urgent priority 1 attenson

Pulse characteristics

Original pitch	280Hz (pitch match)	Pulse spectrum and amplitude weightings	
Harmonicity	Non-harmonic	Frequency	Amplitude weighting
Number of			
harmonics	8	1092	0.89
Duration	100ms	1148	1*
Envelope	Regular	1372	0.794
		1708	0.631
		1932	0.355
		2268	0.25
		2490	0.178
		2828	0.09

Burst characteristics

Number of	
pulses	6
Total length	600ms
Pulse frequencies	280Hz (2 pulses) followed by 390Hz (4 pulses)
Interpulse	
interval	0

Table 14.5   Attenson information: P2(c) least urgent priority 2 attenson

Pulse characteristics.

Original pitch	317Hz (pitch match)	Pulse spectrum and amplitude weightings	
Harmonicity	Non-harmonic	Frequency	Amplitude weighting
Number of			
harmonics	9	703	1.26
Duration	150ms	1268	1*
Envelope	Slow offset	1670	0.5
		1972	0.28
		2114	0.2
		2187	0.2
		2219	0.18
		2241	0.16
		2248	0.16

Burst characteristics

Number of	
pulses	4
Total length	750ms
Pulse frequencies	320Hz (x1) followed by 350Hz (x1), 380Hz (x1) and 220Hz (x1)
Interpulse	
interval	50ms

Table 14.6   Attenson information: P2(b) mid urgent priority 2 attenson

Pulse characteristics

Original pitch	317Hz (pitch match)	Pulse spectrum and amplitude weightings	
Harmonicity	Non-harmonic	Frequency	Amplitude weighting
Number of			
harmonics	9	703	1.26
Duration	150ms	1268	1*
Envelope	Regular	1670	0.5
		1972	0.28
		2114	0.2
		2187	0.2
		2219	0.18
		2241	0.16
		2248	0.16

Burst characteristics

Number of	
pulses	4
Total length	675ms
Pulse frequencies	340Hz (x1) followed by 360Hz (x1), 400Hz (x1) and 220Hz (x1)
Interpulse	
interval	25ms

Table 14.7   Attenson information: P2(a) most urgent priority 2 attenson

Pulse characteristics

Original pitch	317Hz (pitch match)	Pulse spectrum and amplitude weightings	
Harmonicity	Non-harmonic	Frequency	Amplitude weighting
Number of			
harmonics	9	703	1.26
Duration	150ms	1268	1*
Envelope	Regular	1670	0.5
		1972	0.28
		2114	0.2
		2187	0.2
		2219	0.18
		2241	0.16
		2248	0.16

Burst characteristics

Number of	
pulses	4
Total length	600ms
Pulse frequencies	400Hz (x1) followed by 440Hz (x1), 490Hz (x1) and 220Hz (x1)
Interpulse	
interval	0

Table 14.8   Attenson information: P3(c) least urgent priority 3 attenson

Pulse characteristics

Original pitch	262Hz	Pulse spectrum and amplitude weightings	
Harmonicity	Harmonic	Frequency	Amplitude weighting
Number of			
harmonics	8	1028	0.89
Duration	200ms	1310	1*
Envelope	Slow onset	1572	0.56
		1834	0.5
		2096	0.22
		2358	0.18
		2620	0.14
		2882	0.1

Burst characteristics

Number of	
pulses	4
Total length	1000ms
Pulse frequencies	260Hz (x1) followed by 220Hz (x1), 250Hz (x1) and 200Hz (x1)
Interpulse	
interval	50ms, 0, 150ms

Table 14.9   Attenson information: P3(b) mid urgent priority 3 attenson

Pulse characteristics

Original pitch	262Hz	Pulse spectrum and amplitude weightings	
Harmonicity	Harmonic	Frequency	Amplitude weighting
Number of			
harmonics	8	1028	0.89
Duration	150ms	1310	1*
Envelope	Slow onset	1572	0.56
		1834	0.5
		2096	0.22
		2358	0.18
		2620	0.14
		2882	0.1

Burst characteristics

Number of	
pulses	4
Total length	750ms
Pulse frequencies	280Hz (x1) followed by 230Hz (x1), 270Hz (x1) and 220Hz (x1)
Interpulse	
interval	50ms, 0, 100ms

Table 14.10   Attenson information: P3(a) most urgent priority 3 attenson

Pulse characteristics

Original pitch	262Hz	Pulse spectrum and amplitude weightings	
Harmonicity	Harmonic	Frequency	Amplitude weighting
Number of			
harmonics	8	1028	0.89
Duration	150ms	1310	1*
Envelope	Slow onset	1572	0.56
		1834	0.5
		2096	0.22
		2358	0.18
		2620	0.14
		2882	0.1

Burst characteristics

Number of	
pulses	4
Total length	750ms
Pulse frequencies	390Hz (x1) followed by 330Hz (x1), 380Hz (x1) and 310Hz (x1)
Interpulse	
interval	50ms, 0, 100ms

Experimental Validation of the Candidate Attensons

As discussed earlier, the nine candidate attensons were constructed on the basis of prior research knowledge to comply with the requirements that they should vary systematically in perceived urgency and should be recognisable as part of the appropriate priority family of the existing attensons. Group membership and perceived urgency were both tested for in the experiments to validate the attenson design.

Experiment 1: An Investigation into the Perceived Urgency of Nine Candidate Attensons

Introduction In this experiment the perceived urgency of the candidate attensons was assessed using free modulus magnitude estimation (Engen, 1971) – an assessment method which has been shown to be valid for the measurement of perceived urgency (Hellier *et al.*, 1993). The aims of the experiment were twofold. The first aim was to ensure that priority 1 attensons were perceived as being more urgent than priority 2 attensons and that priority 2 attensons were perceived as being more urgent than priority 3 attensons. The second aim of the experiment was to validate the relative perceived urgency of the three candidate attensons constructed for each level of priority in order to find out which were the most and least urgent versions of the attenson for a particular priority. In constructing the attensons their perceived urgency had essentially been predicted from prior research findings (Edworthy *et al.*, 1991, Hellier *et al.*, 1993). On this basis, the predicted order of perceived urgency for the attensons was, from most to least urgent : P1(a), P1(b), P1(c), P2(a), P2(b), P2(c), P3(a), P3(b), P3(c).

Method Two male and 10 female subjects were paid for volunteering to participate in this experiment. All were students or staff from the University of Plymouth, their ages ranging from 19 to 38 years. None of the subjects reported having present or past hearing problems. The stimuli consisted of the nine candidate attensons – three of each priority, as described above. Subjects were run one at a time, while seated in a sound-attenuated booth. They were told the broad nature of the study and were asked to read the following magnitude estimation instructions (adapted from Engen, 1971):

> I am going to present to you, in irregular order, a series of sounds. Your task is to tell me how urgent they are by assigning numbers to them.

When you have heard the first sound, give its urgency a number – any number that you think appropriate. I will then present another sound to you, to which you will also give a number, and a third and so on. Let high numbers represent high urgency and let low numbers represent low urgency. Try to make the ratios between the numbers that you assign the sounds correspond to the ratios between the urgency of the sounds. In other words, try to make the numbers proportional to the urgency of the sound as you hear it. Remember that you can assign any number that you think appropriate. There is no right or wrong answer. I want to know how you judge the urgency of the sounds. Any questions?

The nine attensons were presented twice to each subject. Each subject heard the stimuli in a different random order.

Results and discussion   The mean of all of the subjects' judgments on each stimulus is presented in Table 14.11. All scores are magnitude estimations. P1, P2 and P3 are the candidate attensons for priorities 1, 2 and 3 respectively. The levels a, b, and c for each attenson represent the three versions of each attenson at different levels of urgency.

As is shown in Table 14.11, in general terms, priority 1 attensons were judged more urgent than priority 2 attensons which in turn were judged more urgent than priority 3 attensons. There are a few small exceptions, which will be discussed later. It was also shown that, for each attenson, level (a) was the most urgent level, followed by level (b), and then level (c). The level (c) of each attenson – the least urgent version – was the version that was most similar to the existing attenson. The results also show that, for any particular set of candidate attensons (a, b or c), the priority 1 attenson was judged more urgent than the priority 2 attenson, which in turn was judged more urgent than the priority 3 attenson. This finding is shown more clearly in Table 14.12. These findings together indicate that the attensons for the P1, P2 and P3 attensons were constructed at appropriate levels of urgency relative to one another, and also that three different levels of urgency have been constructed for each attenson.

Table 14.11   The mean magnitude estimate of perceived urgency for each stimulus

Stimulus	P1(c)	P1(b)	P1(a)	P2(c)	P2(b)	P2(a)	P3(c)	P3(b)	P3(a)
Mean	12	15	17.5	7.9	10.8	14	5.8	7	8.8

Table 14.12 **The mean urgency judgments of P1, P2 and P3 for each set of attensons**

	Attenson set		
	c	b	a
Priority 1	12	15	17.5
Priority 2	7.9	10.8	14
Priority 3	5.8	7	8.8

A Spearmans Rank Correlation Coefficient between the predicted and obtained urgency rankings for the nine attensons was 0.967. This provides further support for the claim that the priority 1 attensons were more urgent than the priority 2 attensons which in turn were more urgent than the priority 3 attensons, and also for the claim that the three versions of each attenson differed in urgency within each particular priority. Furthermore, this suggests that the perceived urgency of novel sounds can be successfully predicted on the basis of previous research.

Where the predicted and obtained orders of urgency differed, P2(a) was judged as more urgent than P1(c), and P3(a) was judged as more urgent than P2(c). Thus there was overlap between the least urgent version of P1 and the most urgent version of P2, and between the least urgent version of P2 and the most urgent version of P3. In order to preserve the appropriate level of urgency for priority 1, 2 and 3 attensons, and so aid discriminability between attensons, it was recommended that the same level, (a, b, or c) of attenson for each priority be selected. For example, the selection of P1(a), P2(a) and P3(a) could be recommended.

Experiment 2: An Investigation into the Similarity between the Candidate Attensons and Existing Attensons

Introduction In this experiment the similarity between the nine candidate attensons and the existing attensons was assessed using a similarity rating technique. The aim of this experiment was to ensure that each of the candidate attensons was identifiable as being from the same 'family', or priority, as the existing attenson on which it was modelled. Thus each candidate attenson should be rated as being more similar to the existing attenson from the corresponding priority level than to the existing attensons from other priority levels. For example, any of the three candidate attensons for priority 2 should be rated as being more similar to the existing priority 2 attenson than to either of the existing attensons from priority 1 or 3.

Method   Three male and nine female subjects were paid for volunteering to participate in this experiment. All were students or staff from the University of Plymouth, their ages ranging from 18 to 39 years. None of the subjects reported having present or past hearing problems.

The stimuli consisted of the three existing attensons, from priority 1, priority 2 and priority 3, and the nine candidate attensons – three for each priority. Stimulus pairs were constructed from every possible pairing of the existing attensons with the candidate attensons, so that there were 27 pairs of stimuli. The order of stimulus presentation within each pair of sounds – that is, whether the existing attenson or one of the candidate attensons was presented first, was counterbalanced.

Subjects were run one at a time, while seated in a sound-attenuated booth. They were told the broad nature of the study and were asked to read the following similarity rating instructions (adapted from Loxley, 1991):

> In the following experiment you will hear 54 pairs of sounds, in irregular order. Your task is to rate the similarity of each pair on a scale of 1-9, where nine should be your response if you think the two sounds are almost identical and one should be your response if you think the sounds are totally dissimilar. Thus, you will hear the first pair of sounds and will rate their similarity. When you are ready, I will play the second pair of sounds, then the third and so on. Only use whole numbers when making your ratings. There are no right or wrong answers; it is only your impressions that I am interested in. Any questions?

The 27 pairs of sounds were presented twice to each subject. Every subject heard the stimulus pairs in a different random order.

Results and discussion   The mean of all of the subjects' judgments to each stimulus pairing is presented in Table 14.13. All scores are similarity ratings, where high ratings indicate more similarity. Ex1, Ex2 and Ex3 are the existing attensons for priorities 1, 2 and 3 respectively. P1, P2 and P3 are the candidate attensons for priorities 1, 2 and 3 respectively. The levels a, b and c for each attenson represent the three versions of each attenson at different levels of urgency, from most urgent (a) to least urgent (c).

Table 14.13 shows that the priority 1 candidate attensons were

Table 14.13 **Similarity ratings between all pairs of candidate attensons and existing attensons**

Candidate	Existing attenson		
	Ex1	Ex2	Ex3
P1(c)	8.95	5.29	2.37
P1(b)	7.83	3.58	4
P1(a)	6.08	2.95	1.75
P2(c)	3.88	7.08	4.20
P2(b)	4	6.54	4.04
P2(a)	3.54	6.38	2.95
P3(c)	2.37	2.91	7.75
P3(b)	3	3.38	6.83
P3(a)	3.37	3.70	6.5

judged as being most similar to the existing priority 1 attenson, that the priority 2 candidate attensons were judged as being most similar to the existing priority 2 attenson and that the priority 3 candidate attensons were judged as being most similar to the existing priority 3 attenson. These results indicate that each group of candidate attensons, P1, P2 and P3 were identifiable as being from the same priority group as the existing attenson on which they were modelled. For all three groups of candidate attensons, the least urgent version, version (c), was judged most like the existing attenson, and the most urgent version, version (a), was judged least like the existing attenson.

These findings were further investigated by conducting a 'within subjects' analysis of variance on the similarity ratings, (existing attenson (3) by candidate attenson(9)). There was no significant main effect for the existing attensons, which indicates that there were no overall differences in the similarity ratings given to the existing attensons. The significant main effect for the candidate attensons (F (8,88) = 4.135, $p < 0.0003$) indicated that there were differences in the similarity ratings given to the candidate attensons. A Newman Keuls follow-up procedure revealed that attensons P1(c) and P2(c) received significantly higher similarity ratings than attenson P1(a). What was more interesting, however, was the significant interaction between the existing attensons and the candidate attensons (F (16,176)= 45.949, $p < 0.0001$). This indicated that some candidate attensons were judged as being more similar to some existing attensons than to others, as we had predicted. In particular, the candidate P1 attensons were judged most similar (received higher similarity ratings) to the existing P1 attenson, the candidate P2 attensons were judged most similar to the existing P2 attenson and the candidate P3 attensons were judged most similar to the existing P3 attenson. Thus each of the candidate attensons was recognised as being most similar to the existing attenson on which it was modelled and was thus recognisable as representing the appropriate priority 'family'.

Summary

Taken together, the two experiments provided strong support for the attenson design decisions that were made on the basis of prior research. Experiment 1 indicated that the nine candidate attensons were set at a relative level of urgency that was appropriate to the different priorities, and that three versions of each attenson were constructed which differed in urgency (a–c). Experiment 2 went on to suggest that the nine candidate attensons were identifiable as being from the existing attenson priority group on which they were modelled. They were thus identifiable as being from the same 'family' as the appropriate existing attenson. On the basis of this support, we can conclude that we were successful in creating attensons that were recognisable as being from three distinct levels of priority; that sounded like existing attensons at the same level; and that, within each level of priority, we created three potential attensons, differing in perceived urgency. This not only supports the design objectives of the present candidate attenson set, but also indicates that previous research can be used to predict successfully the perception of novel sounds.

Selecting and Evaluating the Attensons

Here considerations to aid selection of three out of the nine candidate attensons are described in detail. These attensons have been preliminarily screened for two important psychological features: similarity to the existing attensons and perceived urgency. The previous section described the experimental results of this screening in detail. In terms of similarity to the existing attensons, the experimental results revealed that:

- The three priority 1 candidate attensons are similar to the existing priority 1 attenson,
- The three priority 2 candidate attensons are similar to the existing priority 2 attenson,
- The three priority 3 candidate attensons are similar to the existing priority 3 attenson.

The principal differences between the candidate attensons and the existing attensons are that the candidate attensons are shorter, and more urgent.

The nine candidate attensons were also screened for perceived urgency. The experimental finding revealed that the predicted order of urgency and the obtained order was highly correlated. Within each priority, the three candidate attensons varied in their degree of urgency. There was, however, some overlap between the bottom of one category and the top of the next; this is inevitable given the small number of degrees of freedom available in the design process (deriving from the need for some similarity between the candidate and the existing attensons and the short length required of the resultant attensons). Thus the least urgent priority 1 attenson is slightly less urgent than the most urgent priority 2 attenson, and the least urgent priority 2 attenson is slightly less urgent than the most urgent priority 3 attenson. This meant that some constraints needed to be placed on the selection of the subset of three attensons chosen.

Almost all choices of attensons would produce the appropriate gradations of urgency between the categories. A good gradation would be obtained if attensons with a similar urgency letter (a, b or c) were chosen for each category. Thus 1(c), 2(c) and 3(c) would be a good choice. The best selection to maximise urgency differentiation between the priorities would be 1(a), 2(b) and 3(c). Whilst other combinations may be possible, it would be unwise to select an attenson from the bottom end of one category and the top end of the next. For example, the selection of a set which included both 1(c) and 2(a) or 2(c) and 3(a), would not be recommended as urgency differentiation between priorities would not be maximised.

Aside from the requirement for a gradation in urgency between categories, there are a number of other components of an attenson which are required in order to make the attenson psychoacoustically effective. These features are less immediately obvious, but are just as important. Most of these requirements are specified in Patterson's guidelines (1982) and were maintained in the attensons designed in this project.

Among the most important of these recommendations is that the individual attenson pulses have a shaped onset envelope to avoid startle and that they contain several harmonics in the 1–3kHz frequency band. All pulses in the attensons designed in this project have this shaped onset envelope, and they all contain at least six harmonics in the 1–3kHz frequency band. There is also a recommendation that these harmonics be related to ensure that the harmonics are resistant to masking by other noise, and that a fundamental frequency is heard even if this fundamental frequency is not physically present in the attenson. Although the harmonics in the priority 1 and 2 attensons are not related in a simple, integer fashion, they both give a strong sense of pitch. Attensons with this type of harmonic structure are in fact likely to be more urgent, and more resistant to possible masking, than a pulse with harmonics related on an integer basis. Additionally, the region below 1kHz is largely avoided altogether in the attensons presented here, because of the much greater noise level associated with this region. A completely harmonic pulse is likely to sound weak if too many of its lower harmonics are missing. Other than not meeting the requirement for related harmonics – the reasons for which have been explained above – the pulse attensons fully meet the requirements set out in Patterson's guidelines.

The end-users of the candidate attensons recommended that they be shorter, and more urgent, than the existing attensons. The urgency requirement has been addressed in that the lowest urgency attenson in each of the categories is most similar to its equivalent existing attenson, with the other attensons being more urgent. As to the question of length, all of the candidate attensons are shorter than the existing attensons. This was achieved, while retaining the pulse structure of the existing attensons, by using shorter pulse lengths in the candidate attensons.

References

Edworthy, J., Loxley, S. and Dennis, I. (1991) 'Improving auditory warning design: Relationship between warning sound parameters and perceived urgency', *Human Factors*, 32(2), 205–31.

Engen, T. (1971) 'Scaling methods', in J. Kling and L. Riggs (eds), *Experimental Psychology*, London: Methuen & Co. Ltd., 45–86.

Hellier, E., Edworthy, J. and Dennis, I. (1993) 'Improving auditory warning design: Quantifying and predicting the effects of different warning parameters on perceived urgency', *Human Factors*, 35(4), 693–706.

Licklider, J.C.R. (1956) 'Auditory frequency analysis', in C. Cherry (ed.), *Information Theory*, New York: Academic Press.

Loxley, S. (1991) 'An investigation of subjective interpretations of auditory stimuli for the design of trend monitoring sounds', unpublished MPhil dissertation, University of Plymouth.

Patterson, R. (1982) 'Guidelines for auditory warning systems on civil aircraft', CAA, Paper 82017, London: Civil Aviation Authority.

15 Observational Studies of Auditory Warnings on the Intensive Care Unit

CHRISTINA MEREDITH, *Buckinghamshire College, High Wycombe,* **JUDY EDWORTHY,** *University of Plymouth* and **DAVID ROSE,** *University of Plymouth*

Introduction

In environments where there are numerous auditory warnings, such as an Intensive Care Unit (ICU), there are concomitant problems. Many of the warnings that occur are false alarms which may result in the alarms being turned off. Alongside work which looks at auditory warning designs and implementation specifically (for example, Edworthy *et al.*, 1991; Hellier *et al.*, 1993), it is also appropriate to examine in detail what occurs in the environment when existing alarms are activated. This is necessary in order to develop a protocol to test future alarm systems. In an environment such as the ICU, the identification of alarms does not occur in isolation: there are other events happening simultaneously, which may or may not have an impact on future alarm designs.

There have been several studies which have examined problems regarding alarms in critical care areas (for example, O'Carroll, 1986; Kestin *et al.*, 1988; Koski *et al.*, 1990; Momtahan *et al.*, 1993), but these studies do not relate the occurrence of the alarms to the activities either of the patient or of the staff. Stanton (1992) investigated the occurrence of alarms on a coronary care unit (CCU) over a three-day period. In this study it was found that certain alarms (such as the ECG alarm) were directly related to patient activity (such as washing and eating). The responses by the staff to the alarms were also

305

reported; these included checking the patient, changing the infusion and checking the ECG heart rate thresholds. However activities being undertaken by the staff when alarms occurred were not reported.

The present chapter discusses a video study undertaken in the ICU environment to establish the pattern of alarms over an extended period of time. The chapter also discusses some of the problems encountered during the study. The first section discusses some of the issues that should be considered when an observation study is being planned in order to obtain accurate and representative data.

Observation Studies

Observation, either by observers who record the information, or by video camera, is used as a method of collecting overt behavioural data in laboratory, clinical and naturalistic settings. In deciding whether direct observation is the appropriate research method to use for collecting behavioural data, as opposed to other methods, such as controlled laboratory experimentation, standardised tests and questionnaires (Sackett, 1978) and self-report or rating by others (Kerlinger, 1973), there are important issues that need to be considered.

The distinguishing feature of direct observation is that the information is obtained directly rather than through the report of the individual. The experimenter can see what is done, rather than what is said is done and for these reasons it is suggested that direct observation is usually more accurate and of a higher quality than recalled information (for example, Cone, 1982; Smith, 1981). Asking people to report about their behaviour in a specific situation or about their beliefs and attitudes could be inappropriate in some circumstances, as in the case of young children where the questions would probably not be understood (for example Smith, 1981). This type of research is very dependent upon an individual's subjective memory; the questions are usually responded to when the subject is distant from the situation and from the concurrent stresses that influence what they actually do or say (Sackett, 1978). The subject's ability, their willingness to respond and pressure of time may influence how accurately an individual replies to the questions. For example, some familiar, everyday activities that are taken for granted by an individual are often not reported when they are asked about their behaviour in everyday circumstances. Nevertheless these activities would be recorded by an observer (Moser & Kalton, 1971).

Often these factors mean that answers to the questions are unreliable. The fallibility of an individual's memory could cause reported data to be seriously distorted in a way that the direct observation of events, as they occur, would avoid. For example, many people report spending the majority of their time talking, but research has shown that less than 15 per cent of daily activities involves verbal communication (for example, Birdwhistell, 1970). These factors are important when considering an issue such as auditory alarms in the ICU as much of the literature regarding alarms is based on subjective impressions and anecdotal evidence. For example, Schmidt and Baysinger (1986) state that most audible alarms are loud, and produce continuous noxious signals, and that during an emergency it may be more effective to have a problem indicated by a 'pleasant sound'. During the current study, many staff complained that the alarms were 'always going off'.

Sampling Strategies

Often the behaviours or activities of interest that are to be observed do not occur at regular intervals across time, which is pertinent to the occurrence of auditory warnings in an ICU environment. If such activities occur infrequently, it is necessary to have longer observational periods so that these events can then be accurately recorded and so ensure that a representative picture concerning the pattern of behaviours is obtained. Continuous, real-time sampling may be appropriate if, for example, certain alarms occurred at random only two or three times during a 24-hour period.

An important consideration when undertaking an observation study is to ensure that behaviours and activities are represented accurately. Some behaviours may occur frequently and have a long or variable duration, while others are momentary and may have a rapid on/off duration. Other behaviours or activities may occur less frequently and have either a long or a short duration. If a large number of categories are also to be recorded, a combination of the duration of activities described above may occur and real-time, continuous sampling may then be the only way to ensure representative sampling of all the behaviours.

Video Studies

There are many studies that report the advantages of observation using audio or video recording machines over direct observation using observers (for example, Johnson and Pennypacker, 1980). These advantages include obtaining a permanent, accurate and detailed record of the events which can assist memory. With a permanent record such as a video tape, the researcher's interpretation of, for example, particular behaviours, can be confirmed by other investigators who can also analyse the tape. Increasing observer agreement means that the reliability of the study is also increased.

Nevertheless there are disadvantages of using a video camera which are discussed by Weick (1968), who states that, when using certain types of lenses on a video camera, the perspective can be foreshortened; for example, if a group of people are standing far apart, on film they appear to be close together. This particular disadvantage in using these types of lenses may have important implications when considering them for a study in an environment such as the ICU. For example, equipment and monitors may appear very close together and therefore would be difficult to identify when analysing the tape.

Another disadvantage is the time taken to analyse taped data. Typically the field investigation is highly complex and behavioural events that may only occupy minutes in time may take many times that to describe (Barker and Wright, 1955). When analysing the amount of time taken to produce, code and analyse their tapes, Wohlstein and McPhail (1979) found that they had spent 20 hours for each minute of tape. Sackett (1978) also warns that the real danger is spending an inordinate amount of time analysing the tapes. Finally, if recording equipment is used, it can become obtrusive and its presence may have an effect on subject behaviour subsequently.

Reactivity

One major problem with obtrusive observation is that it can be reactive, which means that the behaviour of the participants is influenced by the presence of either observers or the recording equipment (Webb *et al.*, 1981). One example of subject reaction is reported by Belk *et al.*, (1988) who found that, when researchers integrated a video camera into the participant–observer phase of a study, some of the subjects would approach them and ask to be filmed.

It has been suggested that, when people are being observed, they tend to present more positive behaviours perhaps to exhibit a more favourable self-image (Roberts and Renzaglia, 1965). Other studies have postulated that adults become involved in more positive play with children (Zegiob *et al.*, 1975) and participants perform more altruistic acts when being videotaped (Samph, 1969). However, it is impossible to know whether these events would have occurred had there been no observers or video camera present!

Other potential sources of reactivity include conspicuous observers and/or recording equipment, interaction of physical characteristics, for example age or gender of the observer or subject, and the rationale given for undertaking the observation study (Johnson and Bolstad, 1973).

Many ways of minimising reactivity, thereby acquiring an accurate picture of the behaviours of interest, have been suggested. These include the personal appearance and tact of the observers (Haynes, 1978), their professional status (Wallace, 1976) and the relationship, particularly that of trust, that develops with the participants (Weick, 1968). The observers must be able to articulate what they are trying to do and to show a genuine interest in what is going on. It is important to develop a rapport with the participants and an ability to interpret and describe what is observed accurately (Johnson and Pennypacker, 1980).

Other ways of minimising reactivity are suggested by Foster *et al.* (1988), who postulate that participants should agree to be observed but be unaware of the exact times that the observations will occur. The observers should take care to ensure they are not in the way of current activities and encourage the participants to act naturally, which could be achieved by providing an acceptable and non-threatening rationale for the study. The participants should also be reassured that, while the results obtained from the data collected may have important consequences, they will not affect their job.

It is difficult to assess how successful all these tactics are, as there appears to be very little empirical evidence (Foster *et al.*, 1988) to suggest that, if a combination of the above suggestions are used, there will be a positive difference in the results; that is, that the participants will behave naturally. One way to examine whether employing such tactics does improve the results would be to compare the same situation using direct observation and covert, surreptitious recording techniques. However recording without subjects' consent is considered unethical and raises issues such as informed consent of the participants and invasion of the participants' privacy. Studies of this nature are rarely, if at all, undertaken nowadays.

In order to achieve a decrease in reactivity, using any of the suggestions made above, the study should involve an adequate period of investigation and not a quick, superficial visit, as this could result in a seriously biased and unrepresentative picture (Moser and Kalton, 1971). If there are problems with the approach and ability of observers, this could cause participants to feel suspicious (Johnson and Pennypacker, 1980); it could also create effects such as heightened paranoia, hostility and uncertainty, together with changes in the participants' verbal behaviour and behavioural responses (Weick, 1968).

The Current Study

The study reported in this chapter was undertaken as a preliminary investigation into the behavioural responses to alarms and activities that were being undertaken in the ICU when alarms were activated. The study involved the use of a video camera which could record continuously for eight hours, thereby obtaining a substantial amount of data. A single video camera was used to record the equipment and monitors in use at one bed on the ICU, at Derriford Hospital, Plymouth.

In the early stages of the study, a wide-angled lens was used to film all of the four beds in the ward observed. However there were problems in using a wide-angled lens, which not only distorted the size of the four-bedded bay but also obscured the individual detail of the equipment. Another problem transpired when the curtains were pulled around one or more of the beds, which meant that nothing could be observed for those beds. Therefore, to enable a clear view of the equipment, one camera was used with the focus on one bed space.

The criterion used to determine which bed to film was the amount of equipment that was attached to the patient, the patient with most equipment being selected. The video camera was attached to a high stand and positioned inside the curtains so that all the equipment could be seen. This position also ensured that the video camera was relatively unobtrusive and not in the way. The view of the equipment was checked using a small television monitor positioned on a nearby trolley.

Filming was undertaken at random times for continuous eight-hour periods with a timing mechanism used for night-time, late evening and early morning recording. A representative, 24-hour time period was included in the filming, and was obtained over a four-day, non-sequential period. The time periods were as follows: midday to 20.00 hours, 16.00 hours to midnight, midnight to 08.00 hours and 08.00 hours to 16.00 hours. Although there were considerable methodological problems with the video study, nevertheless, as all the sounds on the unit were recorded by the microphone, the data from the video films could be presented as a descriptive 'audio picture' of the auditory warnings and miscellaneous sounds that occurred over the total 32 hours filming in the ICU.

Each video was watched in real time and each time an alarm or a miscellaneous sound occurred three pieces of information were recorded: (1) each alarm was identified, (2) each miscellaneous sound was identified and (3) the onset/offset time of each alarm and each miscellaneous sound was noted. Possible alarms were from the following pieces of equipment: ventilator, pulse oxymeter, syringe pump, infusion pump, electrocardiogram (ECG) and humidifier. The miscellaneous sounds were as follows: telephone, doorbell, doctor's bleep and pulmonary arterial flotation catheter (PAF). The PAF was included under 'miscellaneous' as the sounds recorded occurred when the keyboard was used to type in or request information and a 'beep' sounded each time a key was depressed. The total frequency of the alarms and miscellaneous sounds are shown in Table 15.1.

Although the total numbers of alarms for each session are similar, Figure 15.1 illustrates that the overall frequency of each individual alarm and miscellaneous sound differs across the total duration of the study. It can be seen in Figure 15.1 that the ventilator alarm occurred most frequently (99 times), while in comparison the humidifier occurred the least number of times (five) during the observation study.

However this pattern is not constant for each session recorded during the observation study. Figure 15.2 illustrates the comparison between the totals for each alarm and miscellaneous sound for each session recorded during the observation study. The frequency of alarms and miscellaneous sounds differs for each of the time periods recorded; for example, Day 1 (midday–20.00 hours) the ventilator alarm (34), pulse oxymeter (36) and telephone (31) sounded frequently while the PAF was not heard at all. The ventilator also sounded frequently during Day 2 (16.00–midnight) and Day 4 (08.00–16.00 hours). In comparison, during the recording on Day 3 (midnight–08.00 hours) the most frequent sound occurring was the PAF (66), with the ventilator only occurring eight times.

Discussion

Although one of the aims of the video study was to observe the activities of the staff and their behavioural responses when alarms sounded, this objective was not achieved satisfactorily. However this aim was achieved in a later observational study (Meredith, 1995) in which the relationship between the activation of alarms and behaviour is examined in depth. Several factors such as the unintentional movement of the camera which resulted in the end of a bed or the curtains being filmed for several hours, limited the study reported in this chapter. Although a wide-angled lens was experimented with during the preliminary stages of the study, this

Table 15.1 Total frequency of sounds recorded during the observation study

Day	Midday–20.00 hrs	16.00hrs–midnight	midnight–08.00hrs	08.00hrs–16.00hrs
Total sounds	129	97	110	94

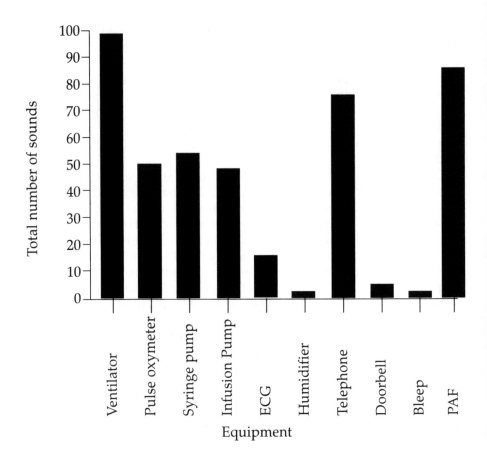

Figure 15.1   **Absolute frequency for all sounds recorded during the observation study**

proved to be unacceptable as the equipment could not be viewed satisfactorily and, as the curtains were often pulled around a patient's bed when a particular procedure necessitated the privacy of the patient, nothing could then be observed for that bed space.

During procedures such as turning the patient, or when drugs were being administered via the syringe or infusion pumps, the view of the equipment was often obscured by the staff as they undertook these procedures, so if staff did respond to alarms the response could not be seen on these occasions. This meant that it was difficult to observe the activities and behavioural responses of the staff when the alarms occurred, although they were visible on some occasions.

On the few occasions when behavioural responses were observed on the tape, it could be seen that the staff immediately silenced the

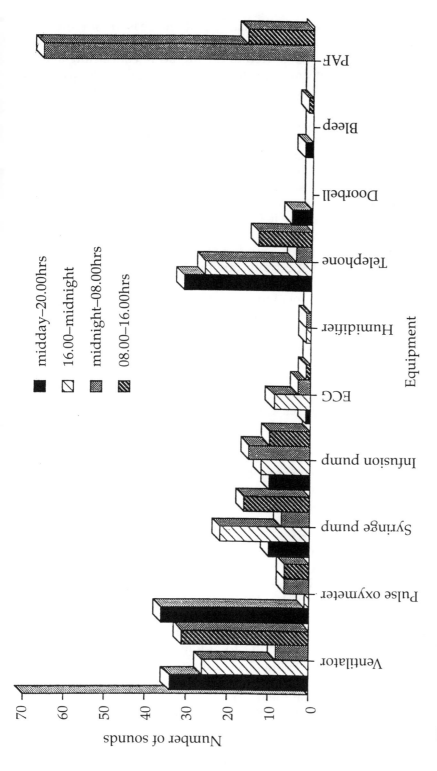

Figure 15.2   Comparisons of the absolute frequency of alarms and miscellaneous sounds occurring during the observation study

alarms. It could be suggested that one of the reasons for the staff silencing the alarms so rapidly is the annoyance caused by many of the sounds, as is widely reported in the literature concerning alarms (for example, Schmidt and Baysinger, 1986; Samuel, 1986). Many of the staff involved in the current study expressed irritation at the disagreeable sound of many of the alarms.

Unwanted noise may also cause distress to the patients (Turner *et al.*, 1975). The psychologically harmful effects of noise on performance are well documented – it could add to the stressors of an already stressful job. Noise may also distract individuals and may mask other important sounds. A non-essential noise such as the 'beep' from the PAF every time data are entered, is an unnecessary addition to the noise on the ICU.

The question as to which alarm occurs appears to depend to some extent on the activities being undertaken by staff. In a similar way Stanton (1992) found that certain activities by patients, such as eating and washing, initiated alarms, for example the ECG alarm. However in the current study the relationship between the activities of the staff and the activation of the alarms could only be inferred, as in the example of physiotherapy and the ventilator alarm. This was because, when activities were being undertaken and alarms were activated, the activities could only be seen clearly on the tape on a few occasions, and consequently the data regarding activities and behavioural responses to the alarms were limited.

There may be various reasons to explain the presence of staff when an alarm is activated. One reason, for example, may be that, if a patient is moving and the electrodes from the ECG monitor fall off, the alarm will sound. This is effectively an artifact – not a true alarm situation. Another reason may be a technical problem, an example being the syringe pump or infusion pump having no drugs or fluid left in them and the staff therefore replacing either the infusion or drugs in the syringe pump.

As discussed earlier in this chapter, the time taken to analyse video data (for example, Sackett, 1978) is often extremely time-consuming. This was certainly true in the current study, as each video was analysed in real time even when there was no appropriate picture on the tape, as the sounds could be clearly heard. Another time-consuming activity involved the immediate identification of a sound; for example, if a sound occurred for as brief a period as one second, it was occasionally difficult to identify it instantaneously. Identification would then entail listening to a prerecorded tape of the sounds from all the various items of equipment (with the spoken name after each sound on the tape).

There is the possibility that the presence of the video camera altered the responses of the staff. For example, they may have answered the alarms more quickly than usual. However the staff did not know when the timer on the video camera was set by the experimenter to start filming (a device suggested by Foster *et al.*, 1988, as a way of minimising reactivity) so it is feasible to infer that the staff did forget about the camera, particularly if they did not know when it was filming. The camera remained on the ward in between filming, so it is also possible that the staff became used to its presence. Feedback from the staff after completion of the study suggests that they did appear to forget about the video. A good rapport was developed with the staff, who were very interested in the rationale for the study and had many positive suggestions and opinions to offer.

While the video study presents a representative 'audio picture' of a non-consecutive 32-hour period, it was not possible to establish which events and activities were occurring on the unit when the alarms sounded, but the study provides useful data regarding alarm frequency nonetheless. For example, it could be inferred from the video that, when the physiotherapist was on the unit, the ventilator alarm sounded considerably more than when physiotherapy was not being undertaken. However, as this relationship could not be directly observed from the tape, it was decided to conduct an observation study in which two observers directly observed the activities of the staff and corresponding alarms on the ICU which can be seen elsewhere (Meredith, 1995).

Other studies have found that video studies can sometimes yield results that are less effective than direct, 'live' observation. For example, Bench *et al.* (1974) and Bench and Wilson (1975, 1976) compared 'live' observation with video recording of neonates and infants and found that little or no information was lost using the former, which they suggest may in fact be superior despite the opportunity for repeated observation available when video recording is used. Sackett (1978) argues that, even in a restricted setting, a video tape can miss much that could be observed live.

References

Barker, R.G. and Wright, H.E. (1955) *Midwest and its Children*, Evanston, IL: Row Peterson.

Belk, S.S., Snell, W.E., Garciafalconi, R., Hernandezsanchez, J.E., Hargrove, L. and Holtzman W.H. (1988) 'Power Strategy Use in the Intimate Relationships of Women and Men from Mexico and the United States', *Personality and Social Psychology Bulletin*, **14**, 439–47.

Bench, J. and Wilson, I. (1975) 'A Comparison of Live and Video Recorded Viewing of Infant Behaviour Under Sound Stimulation, 2. Six-Week-Old Infants', *Developmental Psychobiology*, **8**, 353–74.

Bench, J. and Wilson, I. (1976) 'A Comparison of Live and Video Recorded Viewing of Infant Behaviour Under Sound Stimulation, 3. Six-Month-Old Babies', *Developmental Psychobiology*, **9**, 297–303.

Bench, J., Hoffman, E. and Wilson, I. (1974) 'A Comparison of Live and Video Recorded Viewing of Infant Behaviour Under Sound Stimulation, 1. Neonates', *Developmental Psychobiology*, **7**, 455–64.

Birdwhistell, R. (1970) *Kinetics and Context*, Philadelphia: University of Pennsylvania Press.

Cone, J.D. (1982) 'Validity of Direct Observation Assessment Procedures', in D.P. Hartmann (ed.), *Using Observers to Study Behavior*, San Francisco: Jossey-Bass Inc., 67–79.

Edworthy, J., Loxley, S. and Dennis I. (1991) 'Improving auditory warning design: relationship between warning sound parameters and perceived urgency', *Human Factors*, **33**(2), 205–31.

Foster, S.L., Bell-Dolan, D.J., and Burge, D.A. (1988) 'Behavioral Observation', in A.S. Bellack and M. Hersen (eds), *Behavioral Assessment: a Practical Handbook*, 3rd edn, New York: Pergamon Press.

Haynes, S.N. (1978) *Principles of Behavioral Assessment*, New York: Gardner Press.

Hellier, E.J., Edworthy, J. and Dennis, I. (1993) 'Improving auditory warning design: quantifying and predicting the effects of different warning parameters on perceived urgency', *Human Factors* **35**(4), 693–706.

Johnson, J.M. and Pennypacker, H.S. (1980) *Strategies and Tactics Of Human Behavior Research*, Hillsdale, NJ: Lawrence Erlbaum Associates.

Johnson, S.M. and Bolstad, O.D. (1973) 'Methodological Issues in Naturalistic Observation: Some Problems and Solutions for Field Research', in L.A. Hamerlynck, L.C. Handy and E.J. Mash (eds), *Behavior Change: Methodology, Concepts and Practice*, Champaign, IL: Research Press.

Kerlinger, F.N. (1973) *Foundations of Behavioural Research*, 2nd edn, New York: Holt, Rinehart & Winston.

Kestin, I.G., Miller, B.R. and Lockhart, C.H. (1988) 'Auditory Alarms During Anesthesia Monitoring', *Anaesthesiology*, **69**, 106–9.

Koski, E.M.J., Makivirta, A., Sukuvaara, T. and Aarnck, K. (1990) 'Frequency and Reliability of Alarms in the Monitoring of Cardiac Post Operative Patients', *International Journal of Clinical Monitoring and Computing*, **2**, 47–55.

Meredith, C.S. (1995) 'Auditory Alarms in the Intensive Care Unit: Experimental and Observational Studies', unpublished doctoral dissertation, University of Plymouth.

Momtahan, K.C., Hetu, R. and Tansley, B.W. (1993) 'Audibility and Identification of Auditory Alarms in Operating Rooms and an Intensive Care Unit', *Ergonomics*, **36**, 1159–76.

Moser, C.A. and Kalton, G. (1971) *Survey Methods in Social Investigation*, London: Heinemann Education Books Ltd.

O'Carroll, T.M. (1986) 'Survey Of Alarms in an Intensive Therapy Unit', *Anaesthesia*, **41**, 742–4.

Roberts, R.R. and Renzaglia, G.A. (1965) 'The Influence of Tape Recording on Counselling', *Journal of Counselling Psychology*, **12**, 10–16.

Sackett, G.P. (1978) 'Measurement in Observational Research', in G. Sackett (ed.), *Observing Behavior vol. 11: Data Collection and Analytic Methods*, Baltimore, MD: University Park Press.

Samph, T. (1969) 'The Role of the Observer and the Effect on Teacher Classroom Behavior', *Occasional Papers*, **2**, Pontiac, Michigan: Oakland Schools.

Samuel, S.I. (1986) 'An Alarming Problem', *Anesthesiology*, **62**, 128.

Schmidt, S.I. and Baysinger, C.L. (1986) 'Alarms: Help or Hindrance?', *Anesthesiology*, **64**, 65.

Smith, H.W. (1981) 'Structured Observation', *Strategies of Social Research: The Methodological Imagination*, 2nd edn, Englewood Cliffs, NJ: Prentice-Hall.

Stanton, N.A. (1992) 'Alarms in a Coronary Care Unit', Health and Safety Unit and Applied Psychology, Aston University, Birmingham.

Turner, A., King, C. and Craddock, J. (1975) 'Measuring and Reducing Noise', *Hospitals*, **49**, 85–90.

Wallace, C.J. (1976) 'Assessment of Psychotic Behavior', in M. Hersen and A.S. Bellack (eds), *Behavioral Assessment: a Practical Handbook*, New York: Pergamon.

Webb, E.J., Campbell, D.T., Schwartz, R.D., Sechrest, L. and Grove, J.B. (1981) *Non-Reactive Measures in the Social Sciences*, 2nd edn, Boston: Houghton Mifflin.

Weick, K.E. (1968) 'Systematic Observational Methods', in G. Lindzey and E. Aronson (eds), *The Handbook of Social Psychology*, vol. 2, Melo Park, CA: Addison-Wesley.

Wohlstein, R.T. and McPhail, C. (1979) 'Judging the Presence and Extent of Collective Behaviour from Film Records', *Social Psychology Quarterly*, **42**, 76–81.

Zegiob, L.E., Arnold, S. and Forehand, R. (1975) 'An Examination of Observer Effects in Parent–Child Interactions', *Child Development*, **46**, 509–12.

16 Auditory Alarms in Intensive Care

JOHN WELCH, *King's College, London and University Hospital Lewisham*

Introduction

The modern intensive care unit (ICU) contains increasingly numerous, varied and sophisticated equipment for the support and monitoring of critically ill patients. There has been a concomitant increase in the warning systems designed to signal changes or problems in equipment functioning, and variations in patients' conditions.

Auditory alarms are most often used in the ICU 'because hearing is a primary warning sense' (Patterson, 1990) and has 'generally omni-directional characteristics' (Wilkins, 1981). Auditory alarms can be detected even when staff are not specifically focused on monitoring, and may be attending to other complex tasks. In contrast, visual warnings may be detected only if they occur within the visual field. Verbal alarms are also unsuitable in the clinical setting: they are unethical in that vulnerable patients and their visitors might hear highly personal, specific and potentially very frightening messages about other patients, and because verbal warnings may be easily lost amongst other, non-critical, verbal information (Edworthy, 1994).

However, it has been recognised for some time that the proliferation of auditory alarms in ICU is not without its problems (for example, Kerr, 1985; Meredith and Edworthy, 1994). 'Proper design and implementation is often found wanting' (Meredith and Edworthy, 1994), and there is a tendency to install warnings on a 'better safe than sorry' basis (Patterson *et al.*, 1986). ICU auditory alarms are often too numerous, confusing, too loud (or sometimes easily masked), too startling, harsh and intrusive to enable easy identification and response (Patterson *et al.*, 1986; Momtahan *et al.*, 1993). Furthermore, there is no coherent approach to the pairing of ICU warnings, in terms of their perceived urgency, with the medical significance of the events that they indicate (Momtahan and Tansley, 1989; Finley and Cohen, 1991). Even alarms on machines made by the same manufacturer are not always arranged in any obvious system, and coordination between makers is rare (Meredith and Edworthy, 1994). Thus warnings on two machines that perform the same function, but are made by different companies, often sound different; while two similar sounding alarms may signal different problems.

Too Many, Confusing Warnings

A typical, *stable* intensive care patient is attached to a mechanical ventilator, physiological monitors – electrocardiograph (ECG), arterial blood pressure, central venous pressure, blood oxygen saturation, central and peripheral temperatures – and a number of intravenous infusion (IVI) control devices. If he or she is seriously ill, extra monitors, more IVI devices, kidney replacement therapy equipment, and perhaps a special pressure-relieving bed are likely to be added. In the worst cases, further technological interventions are not uncommon. All of these items emit one or more auditory alarms in various situations, so that there may well be over 30 different warnings associated with a single intensive care patient (Edworthy, 1994). Each patient in each bed is able to trigger several alarms simultaneously, and ICUs generally have between four and eight beds in one space, with a consequent potential for many warnings to sound at the same time.

The likelihood of confusion has been illustrated by several studies. Momtahan *et al.* (1993) asked ICU nurses to identify tape-recordings of the 23 auditory alarms from their unit. The nurses were allowed to take their time, replay the tapes and adjust the volume as they pleased, yet could correctly identify only 14 sounds on average (61 per cent), even on a 'liberal' identification criterion. They scored a mean of nine hits (39 per cent) on a more stringent criterion. Similarly, Loeb *et al.* (1990) found that anaesthetic clinicians correctly recognised just 34 per cent of 19 warnings from their own operating theatres. Subjects heard each alarm separately, and it can be assumed that they gave their full attention to the task. In real life, identification is facilitated by sound localisation and various visual cues. However, staff also have their attention divided by a wide range of physical and mental demands, and often have to process two or more warnings at once. Auditory alarm identification in clinical situations is likely to be no better than laboratory performance. It should also be noted that many warnings consist of high frequency sine tones that are difficult to localise (Edworthy and Meredith, 1995); and that accurate memory of this kind of sound rapidly decays (Deutsch, 1978).

The majority of staff tested in these studies heard the alarms many times during the course of their work – O'Carroll (1986) recorded 1455 warning soundings in a relatively quiet ICU over a three-week period – yet they could correctly recognise only one-third of the auditory alarms from their units. It may be that learning is a problem because some of the warnings sound very similar (for example, Meredith and Edworthy, 1994), because alarms can often mask each other if triggered simultaneously (Momtahan *et al.*, 1993); and also because there is no correlation between the perceived urgency of warnings and the events that they mark (Momtahan and Tansley, 1989; Finley and Cohen, 1991). But all of these factors are made worse by there being too many alarms in the first place. When McIntyre (1985) used a questionnaire to ask 852 anaesthetists how many individually recognisable auditory warnings they thought that they could 'cope with', 87 per cent said that six or fewer alarms would be best, with five being the preferred number.

The difficulty of accurate recognition of auditory warnings in the ICU is known to have had fatal consequences; for instance, Cooper and Couvillon (1983) report a patient death attributed to confusion between the ventilator and ECG alarms.

Aversive, Inappropriate Qualities of Warnings

Patterson (1982) described the behaviours of flight crews responding to loud auditory alarms in aircraft. The crews' attention was often directed to locating and cancelling the warnings in question, rather than to identifying and addressing the actual causative problem. He suggested that manufacturers may make alarms excessively loud and strident to ensure that they command attention, without giving due consideration to the disruptive effects on thought processes and intra-team communication. Nonetheless, he recommended that auditory warnings be set at a volume of 15–25dB above background noise (the masked threshold). In a health care setting, Falk & Woods (1973) measured noise produced by hospital staff and patients of up to 86dB(A), with average levels of 55–60dB(A). The superimposition of a number of alarms at 15–25dB above this kind of background noise would then result in highly annoying peak noise levels of *at least* 70dB(A).

Ultimately, clinicians themselves identify problems with the current systems. For example, when anaesthetists working in operating theatres were asked about their interactions with auditory warnings by McIntyre (1985), although most said that alarms have a useful role in patient management, the majority also reported that they sometimes deliberately deactivated warnings, even at the very beginning of a case. The anaesthetists gave various reasons for disconnecting the devices: there were too many false alarms; for 'peace and quiet'; because the sound itself was disliked; because of confusion between different warnings; and because the signals hindered proper attention to the patient.

The false alarm rate in ICU was highlighted by O'Carroll (1986). Of 1455 warning soundings in three weeks, only eight marked potentially life-threatening events. The great majority of ventilator alarms were artifactual, triggered during airway suction, physiotherapy or by patient movement, while most monitor warnings were due to displacement of ECG leads, usually by patient movement. A similarly high level of specious signals was noted by Kestin *et al.* (1988), during their observation of 50 surgical operations. They found that an alarm sounded on average every 4.5 minutes, but 75 per cent of warnings were 'spurious', caused by patient movement, interference or mechanical problems. Only 3 per cent of alarms signalled risk to the patient. Undoubtedly some of the warnings gave useful information: for instance, signalling the completion of particular intravenous infusions. However, when so few alarms actually indicate critical situations, watchfulness for such events tends to be attenuated. Indeed, Bliss *et al.* (1995) showed that response to a particular signal is closely associated with its perceived reliability – that is, a high proportion of false alarms will tend to reduce the response rate. Furthermore, because there is no systematic association of important sounding warnings with significant problems, staff cannot choose to concentrate attention solely on urgent sounding alarms without risking missing an emergency signalled by an innocuous sound.

The majority of auditory warnings emitted in an ICU are in fact false alarms, triggered by interference or 'noise'. But it is also suggested that the normal operating band of the warning devices is too narrow, so that relatively small and normal changes in physiological functioning are signalled (for example, Edworthy and Meredith, 1994). Monitors are generally equipped with both preset and adjustable alarm thresholds, but limits are usually set within very conservative boundaries. There is too often little consideration of individual differences, such as what is normal and abnormal for a particular patient, rather than for an average healthy young adult.

Poor 'Urgency Mapping'

The problem of poor 'urgency mapping' between the perceived urgency of auditory warnings and the actual urgency of the situation indicated is well documented (for example, Momtahan and Tansley, 1989; Edworthy, 1994). Finley and Cohen (1991) asked subjects to rate the perceived urgency of various hospital auditory alarms, and correlated the ranked ratings with ranked ratings of the urgency of the clinical events indicated by the warnings. Half the alarms had a perceived urgency that was significantly different from the actual importance of the situation that they signalled. One warning was given the highest perceived urgency rating, but was linked to an event of so little importance that the authors recommended that it should not have an auditory alarm at all, and rather should be indicated by a visual warning. The disparity between perceived or psychoacoustic urgency of alarms and the urgency of the medical situations that they signal would be of less significance if the meanings of the warnings were well known and easily identified, but this is clearly not the case. If signals and events were matched for urgency, staff would have a basis for prioritising their responses, even if the meanings of the alarms were not known. This would be especially useful should several warnings sound at once.

Other Considerations

The work summarised above highlights the problems facing operators of medical equipment and their associated auditory alarms. It should also be pointed out that the systems and practices in current use contribute significantly to noise in the ICU, and have a number of deleterious effects on hospitals' clients: that is, patients and their visitors. There is also evidence of other, wider implications for ICU personnel.

Effects of Auditory Warnings on Patients

Many intensive care patients are heavily sedated, may be effectively paralysed with muscle relaxant drugs, or may be already comatosed because of drug overdose, trauma or severe illness. These patients have no way of controlling their environment or directly communicating their needs, but are still passive receivers of 'a massive array of sensory stimuli' in an 'atmosphere ... not unlike that of the tension-charged war bunker' (Hay and Oken, 1972). The patients' hearing may remain intact in these conditions, or is perhaps heightened; yet tolerance of noise is lowered in sickness (Turner *et al.*, 1975). Noise is a considerable factor in the 'sensory overload' that can have highly stressful physiological and psychological consequences, such as sympathetic nervous stimulation, sleep deprivation, changes in affect, cognitive impairment, and even 'critical care psychiatric syndrome' (Noble, 1979). It is believed that critically ill patients develop a certain egocentricity, tending to interpret *all* stimuli as directly relevant to themselves (Smith, 1990). ICU beds are often arranged in relatively close proximity, so that it is difficult to tell the source of many sounds, especially auditory alarms. Therefore the individual ICU patient hears warnings from different beds, but may think that they pertain to him or her only.

Bentley *et al.* (1977) found daytime noise in a London ICU to be equivalent to heavy street traffic, with a loud noise of >70 dB(A) occurring every two minutes. Redding *et al.* (1977) measured auditory alarms at an ICU patient's head at 71 and 92dB(A) (heart rate and ventilator disconnect warnings respectively). Individual responses to noise are somewhat subjective and variable, but a more recent review indicates that many people find noise over 60dB(A) 'annoying' (Baker, 1993); while it is claimed that noise levels less than 35dB(A) are desirable for sleep and its important healing function (Adam and Oswald, 1984). Sounds that are higher-pitched, harsh-toned, and have a sudden onset with a sharp peak are perceived as particularly annoying (Baker, 1993); and are the very characteristics of many auditory signals.

Another related problem is that many ICU patients are given drugs with potential side-effects of deafness, especially when taken in an environment with continuous supranormal noise. Falk and Woods (1973) found hearing loss in patients given these drugs at a noise level of 58dB(A) in an acute care unit where *average* noise levels over twenty-four hours were 55–60dB(A).

It has also been suggested that excessive noise depresses pain thresholds: Minckley (1968) reported that patients' requests for analgesia increased as noise rose from 60–70 dB(A). This could be particularly significant for ICU patients suffering painful injuries, disease and/or surgery, but often unable to make their distress known.

Effects on Patients' Visitors

Visitors to patients in the ICU are understandably anxious and often further discomforted by the strangeness of the surroundings. Most sit passively and quietly at the bedside, closely watching the visual displays and listening for alarms. Fearful startle reactions may be observed whenever an auditory warning sounds, even though the alarm is most likely to be connected with another patient and, as O'Carroll (1986) showed, rarely indicates a critical event. Staff strive to reassure and keep visitors well informed, but the ICU environment itself, including the warning systems, militates against this.

Effects on Staff

ICU staff themselves find their work environment noisy and stressful, as illustrated by the anaesthetists in McIntyre's (1985) study, who disabled alarms partly because they found them annoying and distracting. Topf and Dillon (1988) used hierarchical multiple regression analyses to isolate the effect of noise on critical care nurses from two hospitals. They found a highly significant correlation between noise-induced occupational stress and staff burnout. The sounds rated as most stressful were 'continuous beeping of patient monitoring devices' and equipment auditory warnings. When many of the signals are in fact false alarms or indicate low-priority situations, it may be that nurses perceive the warnings as unnecessary at some level. However, many of the alarms are so aversive that a startle response is still evoked, even if the meaning is 'known' to be of relatively little significance. Such a response is ecologically valid, but the illogical pairing of warnings and events in ICU generates frequent incidents of unwarranted arousal. This simply serves to make the alarms more irritating.

Focus of the Present Study

Clinicians and researchers are agreed that auditory warnings in ICU are useful but problematic aids for safeguarding patients. However, the current designs and haphazard arrangements of alarms are largely unsatisfactory, and it often seems that a lot of noise is produced for relatively little useful effect. Although strenuous efforts are being made at domestic, European and international levels to formulate ergonomic standards, there is as yet no real improvement in the day-to-day situation in the British ICU. Indeed, there is a continued growth in the number of machines and associated warnings used to support and monitor the critically ill. In the past, UK ICUs have generally followed the Intensive Care Society (1983) guidelines that recommend at least one nurse to care for each patient. But there are increasing pressures to cut staffing levels in ICU. This means that fewer nurses are watching and listening for more alarms, as well as performing myriad other roles. Not surprisingly, Meredith and Edworthy (1994) predict that, as workload increases, so will auditory confusion.

It has been acknowledged that ICU auditory warnings are often confusing, inappropriate and, especially, too numerous. Momtahan *et al.* (1993) highlight that implicit in this concern is the assumption that it is impossible to learn and remember the meaning of so many alarms. Several studies cited above have shown that intensive care and operating theatre staff can identify only one-third of the warnings recorded in their own work-areas (Loeb *et al.*, 1990; Finley and Cohen, 1991; Momtahan *et al.* 1993). McIntyre (1985) reported that anaesthetists would prefer to have just five alarms to deal with, while Patterson (1990) recommended that there should be no more than about six 'immediate-action' warning sounds. However, less research has been done on the actual learning and retention of auditory alarms, and the present author was unable to find any reports detailing the learning of hospital warnings.

Patterson (1982) described laboratory work showing that it is much more difficult to learn a set of arbitrary auditory stimuli than a list of words. If auditory signals linked with a specific meaning are to be learnt, subjects must learn the sounds themselves, as items of a set to be remembered, as well as the associated events. Patterson and Milroy (1980) examined how quickly subjects could learn and subsequently identify a set of 10 aircraft auditory alarms. (They selected 10 from a possible 54 sounds to maximise acoustic distinctiveness.) Trials consisted of a presentation session when each warning in turn was heard paired with a visual display of its name, and a test session when the alarms were played without their labels. Subjects were then required to give the initial letter of the stimulus name. A list of the warning names was on continuous display, so that subjects could refer to them throughout. They learnt four to six alarms after four trials, but then found committing new warnings to memory more difficult: nine further trials were required to learn two more items. However, all subjects eventually learnt all 10 alarms. One week later, a test showed that they had retained seven of the 10 warnings.

The present study reports two experiments designed to investigate aspects of the learning, retention and confusion of ICU alarms. Experiment 1 tested experienced ICU nurses' identification of auditory alarms recorded in their own unit, taking into account the degree of urgency-mapping of warnings to events as well as the distinguishing qualities, or distinctiveness, of the sounds; and the amount of prior exposure to these signals. Learning and retaining the meaning of alarms can be considered as a particular kind of paired associate learning and, in a comprehensive review, Goss and Nodine (1965) identified both distinctiveness and familiarity of stimuli as key factors in such learning. To the present author's knowledge, auditory warnings had not previously been specifically investigated within this particular paradigm, but it is likely that some of the same principles apply.

In Experiment 2, nurses in training learnt the meanings of ICU auditory alarms in two conditions. This was to determine whether or not urgency-mapping signals and particular causative problems might promote better learning and identification of warnings in the ICU, in addition to being a 'common-sense' component of sensible alarm design, as suggested above. It was hypothesised that, if signals were urgency mapped to similarly urgent situations, hearing a particular warning could help concentrate the search for its meaning amongst events more or less correlated in importance. Thus one group heard alarms urgency mapped with a set of clinical events, while a second group heard a reverse arrangement – that is, the most urgent warnings paired with the least significant situations and vice versa. The second condition is analogous to the current real-life situation in intensive care.

The project is a preliminary exploration of the learning of very specific material in a highly specialised field. There has been much research into the theory and practice of learning, but little work on the problems of learning auditory alarms, particularly in the hospital setting. It would be of considerable interest and value to discover how the learning of warnings could be optimised – for example, how the training of personnel should be organised and alarm systems rationalised to facilitate the effective use of such systems in patient care. There are many potential interacting influences on performance in complex settings like the ICU, and it is beyond the scope of the study to address all the issues involved. However it is hoped that some of the variables influencing learning, retention and confusion of warnings in ICU will be highlighted, with the goal of suggesting methods of improving ergonomic performance in this important and problematic area.

Experiment 1: Identifying Auditory Warnings in the ICU: Method

Design

Experiment 1 had a 'within subjects' 2 x 2 x 2 design, giving eight conditions. A total of 10 auditory alarms recorded in the subjects' workplace were used as stimuli. Six different warnings matched the characteristics of six of the experimental cells, while the two remaining conditions were each represented by two alarms, so that the effects of these could be averaged on analysis. (The researcher recognised that the two cells with two stimulus warnings should be checked for intra-cell variance: see 'Results' below.)

Three independent variables, each at two levels (high and low), of the stimulus alarms characteristics were investigated:

1 psychoacoustic distinctiveness – 'high distinctiveness' warnings taken from the top half of a list of alarms ranked on distinctiveness, versus 'low distinctiveness' warnings from the bottom half of the list;
2 previous exposure to each alarm – that is, signals often heard during ICU activity, versus those infrequently heard;
3 the urgency mapping of the warning to its associated event – that

is, 'highly urgency-mapped (UM)' paired associates well correlated on perceived urgency (whatever their absolute urgency rating), versus 'low UM' paired associates that are poorly or negatively correlated.

The dependent variable was the number of errors made in identifying each warning.

Recording, and Rating of Auditory Stimuli and Medical Events

Twelve (other) subjects rated the alarms for perceived urgency using a modulus magnitude estimation method recommended by Hellier *et al.* (1995), after Engen (1971). The participants comprised 11 females and one male with no significant experience of hospital warnings, so that they could judge the sounds solely on their psychoacoustic properties. They were aged between 24 and 36 years (mean age, 29 years) and had normal hearing. They listened to six differently randomly ordered sets of 15 alarms that were first recorded – three practice sets and three sets that were subsequently scored, an approach which Hellier *et al.* (1995) assert gives the most reliable results. The participants were required to rate the urgency of the sounds by ascribing numbers to them, with high numbers signifying greater urgency and smaller numbers less urgency. They themselves chose to use a 1–10 scale. Subjects were asked to compare each warning's urgency with the others heard, and to try to use values consistent with their relative urgency. The perceived urgency ratings were averaged across all the responses, and the alarms then ranked accordingly. Ten of these 15 sounds were selected to be the experimental stimulus items as described above.

The subjects also judged the distinctiveness of the signals, using the same method. The alarms were then ranked from most to least distinctive. The frequency of warnings sounding in the ICU was also measured by the author in six half-hour sessions, during day and night, over a one-month period.

A selection of clinical situations that might be indicated by an auditory warning were rated in a similar way by 15 expert ICU clinicians comprising five males and 10 females aged 25 to 42 (mean age, 32 years). The event ratings were scored, ranked, and 10 well-demarcated medical situations were chosen to be paired with the auditory stimuli.

None of the subjects who estimated the perceived urgency (and distinctiveness) of alarms or clinical events took part in the two experiments detailed in this report.

Experimental subjects

The subjects of Experiment 1 were trained intensive care nurses employed in a busy six-bed central London ICU. They volunteered for the study. The participants had worked in the health authority's ICUs for between four months and five years (mean experience, two years and two months). They comprised 11 females and two males, aged between 24 and 33 years (mean age, 28 years). All subjects reported good hearing.

Auditory Stimuli: Apparatus

Subjects listened to three differently randomly ordered stimulus sets of 10 different alarms recorded in their unit. Individual signals in a set lasted 10 seconds, with intervals of 10 seconds between each. Subjects heard the stimulus tapes played on an Amstrad MCD750 stereo double cassette player. They had written instructions and response sheets.

Procedure

The nurses were tested in a fully equipped but unused side-room of the ICU. They listened to three differently randomly ordered sets of 10 alarms and were required to respond to each sound with either the name of the machine that emitted the warning or the particular patient/equipment problem that would trigger the signal.

Experiment 1: Results

There was an average error in ICU auditory alarm identification of 37 per cent for all subjects over all trials of the various stimuli. The mean error for the eight conditions is shown below in Table 16.1 with by far the most error caused by signals rated 'low' for both urgency mapping and distinctiveness. Six cells were represented by one stimulus warning each, and two cells were given the averaged effects of two alarms (see 'Design' above). There was no significant difference between the two error scores obtained for each of these two conditions.

The Main Effects of Distinctiveness, Exposure and Urgency Mapping of Alarms

A 2 x 2 x 2 completely randomised analysis of variance yielded the following results. There was a significant main effect of

Table 16.1   Error in identifying ICU warnings by ICU nurses (percentage of total possible error) in high and low conditions or urgency mapping, previous exposure and distinctiveness (*n* = 13)

High Urgency Mapping		Low Urgency Mapping	
High Distinctiveness 28.1	38.2	High Distinctiveness 20.5	3.8
Low Distinctiveness 12.8	16.5	Low Distinctiveness 82.6	71.4
High prior exposure	Low prior exposure	High prior exposure	Low prior exposure

distinctiveness of stimuli: $F(1,12) = 37.42$ ($MSe = 372.45$), $p < 0.001$. Alarm identification error in high distinctiveness conditions was 22.7 per cent on average, but doubled in low distinctiveness conditions (45.8 per cent). There was also a significant main effect of urgency-mapping of warnings and their meanings: $F(1,12) = 10.75$ ($MSe = 1035.79$), $p < 0.01$. The high UM conditions mean error was 23.9 per cent, but low urgency mapping gave a 44.6 per cent error. There was no significant effect of prior exposure to stimuli. The main effects of the three independent variables on warning identification are shown below in Table 16.2 and displayed graphically in Figure 16.1.

Table 16.2   Summary of error in identifying ICU warnings by ICU nurses (percentage of total possible error): main effects of distinctiveness, exposure and urgency mapping (*n* = 13)

	High	Low	Significance of difference
Distinctiveness	22.7	45.8	$p < 0.001$*
Exposure	36.0	32.5	$p > 0.05$
Urgency mapping	23.9	44.6	$p < 0.01$*

*highly significant difference.

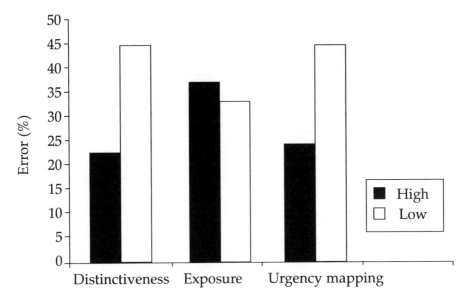

Figure 16.1 **Main effects of high and low levels of distinctiveness, exposure and urgency mapping on warning identification in the ICU**

Interactions Between Variables

Two significant simple interactions were found. The distinctiveness and urgency mapping variables interacted: F (1,12) = 81.69 (*MSe* = 552.24), p < 0.001; this is displayed graphically in Figure 16.2 below. Overall, most error was caused by warnings that were low UM and least distinctive, but a simple effects analysis revealed that a highly distinctive alarm actually produced *more* errors than a less distinctive signal when both were well urgency mapped with events: F (1,12) = 25.72 (*MSe* = 172.6), p < 0.001. However, when warnings were not urgency mapped, a distinctive sound gave far fewer errors than a less distinctive one: F (1,12) = 72.63 (*MSe* = 751.73), p < 0.001.

Frequency of exposure and urgency mapping also interacted: F (1,12) = 4.78 (*MSe* = 594.96), p < 0.05. There was no significant difference in identification of high and low exposure alarms when both were urgency mapped, but a high exposure/low UM warning caused more errors than a low exposure/low UM alarm: F (1,12) = 8.54 (*MSe* = 296.69), p < 0.025. This is shown in Figure 16.3 below.

There was no significant interaction between distinctiveness and prior exposure to alarms and no significant three-way interaction between the variables.

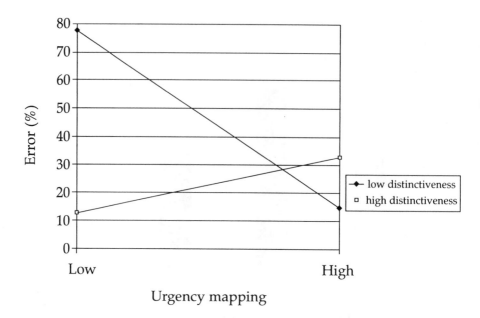

Figure 16.2 Alarm distinctiveness and urgency mapping interaction

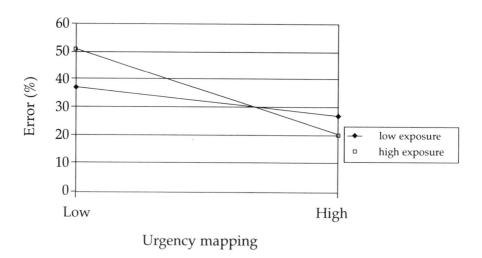

Figure 16.3 Exposure to alarms and urgency mapping interaction

Experiment 2: Learning Auditory Alarms (Method)

Design

Experiment 2 had a 'between subjects' design. There was one independent variable, at two levels – whether the to-be-remembered (TBR) pairs of auditory warnings and medical events were urgency mapped, or not. The dependent variable was the number of alarm identification errors made at various test points during the procedure. It was predicted that subjects learning urgency mapped (UM) warnings and events would make fewer errors than subjects learning unmapped pairs.

Subjects

The subjects were 37 nursing students in their third year at two London teaching hospitals. All were volunteers. They comprised eight males and 29 females, aged between 21 and 34 years (mean age, 25 years). All the subjects reported good hearing and had a satisfactory understanding of the nature and priority of the clinical conditions used to label the stimulus sounds, but little or no experience of using apparatus equipped with alarms (that is, they knew few warning/event associations). Subjects were randomly allocated to the two experimental groups: the urgency mapped group ($n = 18$); and non-urgency mapped group ($n = 19$).

Auditory Stimuli

A total of 15 auditory alarms were recorded initially, from a range of equipment commonly used in ICU: respiratory and renal support machines, physiological monitors, IVI and feed control devices, and a bed. The sounds were recorded at 100cm distance, level with the source, on to Maxwell Capsule UDX II audio-tape cassettes using a Sony ECM-202 Electret condenser microphone (mounted on a boom) and a Sony Walkman WM-DC6 Professional stereo cassette-recorder. The warnings were standardised on length, right/left channel balance and volume on a system comprising an Akai CS-F11 stereo cassette deck, a Sound Lab VME 605 stereographic equaliser-mixer and a Technics RS-BX606 stereo cassette deck. Ten of the 15 alarms were finally selected to be experimental stimulus items (for both experiments 1 and 2) by virtue of their being well differentiated on a scale of perceived urgency. The 10 warnings were differently randomly ordered in a number of stimulus sets, with sets comprising either seven or 10 alarms. Individual signals in a set lasted 10 seconds, with intervals of 10 seconds between each.

The warnings were each given an accompanying name (different from their 'real life' identities) taken from a list of medical conditions rated for significance. In the UM condition, alarms ranked from most urgent sounding to least urgent sounding were paired one-to-one with names of events ranked for clinical importance; that is, the most attention provoking warning was paired with 'cardiac arrest' (the most critical incident) and so on. In the non-UM condition, the first five perceived urgent alarms were randomly paired with the bottom five ranked situations, and the lower five perceived urgent warnings coupled with the top five events. This arrangement was made to contrast with the UM condition without giving a systematic, negatively correlated association between signals and incidents; and to mimic the position in the ICU.

Rating of Auditory Stimuli and Medical Events

The processes of rating of the alarms' psychoacoustic perceived urgency and the rating of the relative significance of clinical situations are described above, in the account of Experiment 1.

Apparatus

Subjects heard the stimulus tapes played on an Amstrad MCD750 stereo double cassette player. They were supplied with written instructions and response sheets for use at test.

Procedure

First session   Subjects were informed that they were participating in a study 'designed to see how best the meanings of auditory warnings are to be learnt'. They heard one set of sounds (without their identifying labels) to allow withdrawal from the exercise should the alarms prove too annoying. None of the subjects left. The researcher told the subjects the events to be learnt in conjunction with the warnings, reinforced by a visual display of the names that was removed after two minutes. Subjects were asked to think about the qualities of each sound as they heard it, and to try to integrate the sound into an imagined scene of themselves reacting to each event as it was identified; for instance, responding to a patient call-button. This instruction was given to encourage a uniform approach to learning. It may also be somewhat akin to the interactive process of learning alarms during work in the ICU. Subjects then heard two sets of (the same) seven warnings, with each alarm named immediately after it sounded. Sets of seven were chosen because previous research (for example, Patterson and Milroy, 1980) and a pilot study prior to the present study indicated that subjects found it difficult to learn more than five or six sounds in a few trials. The identifying labels were either urgency mapped with a particular stimulus, or not, depending on the condition. A third playing of the signals acted as the baseline Test 1. Subjects were then told the correct names of the warnings in Test 1, before hearing two more learning trials, followed by Test 2. The session took approximately 35 minutes.

Second session   Five days later, subjects were tested for their memory of the seven alarms, and given feedback as before. (They had not been informed that their retention would be examined.) They were then told that there would be two more learning trials, with three new warning/event pairs added to the seven previously heard – to make sets of 10 sounds. This was to simulate the real-life situation in which new equipment and associated alarms are frequently introduced. Finally, Test 4 examined identification of all the warnings. As a postscript, subjects were asked to note examples of the alarm/event integrations which they had used to remember the pairs. They were then debriefed. The second session lasted 30 minutes.

Experiment 2: Results

A number of subjects in both conditions made no errors or only one error in auditory warning identification after two learning trials. It seemed unlikely that the experimental manipulation had made any real difference at this early stage, so these subjects' scores were disregarded. Six UM subjects and three non UM subjects were lost. It should be noted that there was no significant difference between the two groups, with or without the discarded subjects, at the baseline Test after two trials; for example, unrelated $t = 1.09$, $df = 26$, $p = 0.286$, two-tailed test ($n = 12, 16$). This left 12 subjects in the UM group and 16 in the non-UM group who could be said to be open to learning during the procedure, and whose scores were subsequently compared. The mean alarm identification error for each group on the four tests is shown below in Table 16.3, and displayed graphically in Figure 16.4. The figures are errors expressed as percentages of total possible error on a given test. Tests 1, 2 and 3 examined recall of seven TBR pairs. Test 4 required 10 responses.

There was only one significant difference in performance between the two groups over the whole exercise: following five learning trials during the first session (that is, at Test 2), subjects learning urgency mapped warnings and events made significantly fewer errors than those learning unmapped pairs: unrelated $t = 2.18$, $df = 26$, $p < 0.025$, one tailed test. The UM group achieved 5.7 hits/1.3 misses.

Table 16.3 **Error in learning urgency mapped alarms, versus learning non-urgency mapped alarms over two sessions separated by a five-day interval (as percentage of total possible error)**

	First session			Second session	
	Test 1	*Test 2*	*5-day interval*	*Test 3*	*Test 4*
	'Baseline' test after 2 learning trials	After 5 learning trials		Testing recall after 5 days	After 3 more trials; 2 with added TBR pairs
	7 TBR pairs	7 TBR pairs		7 TBR pairs	10 TBR pairs
*Urgency mapped** group Mean error (sd: 17) %	53.5	19.1 (sd: 18.7)		58.2 (sd: 25.2)	28.3 (sd: 21.7)
*Non-urgency mapped group*** Mean error (sd: 22.4) %	61.6	37.6 (sd: 26.1)		53.6 (sd: 24.8)	31.2 (sd: 19.6)

*n = 12
**n = 16

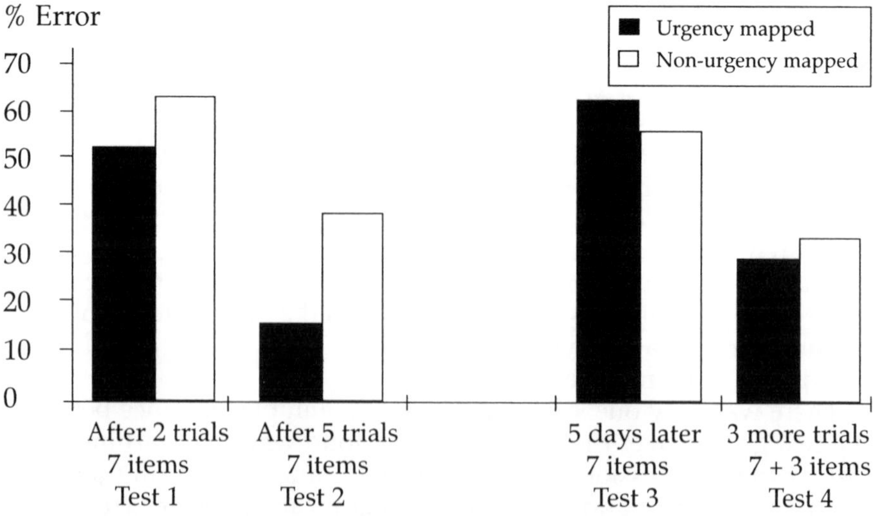

Figure 16.4   Error in learning auditory alarms

Nonetheless, even the non-UM group had learnt a significant number of alarms by Test 2: related $t = 4.06$, $df = 15$, $p < 0.005$, one tailed test – that is, 4.4 hits/2.6 misses.

However, after a five-day interval at Test 3, all subjects had regressed to the baseline error rate measured at Test 1, with no significant difference between the groups. Following feedback on Test 3, subjects had three further learning trials, with the original seven TBR pairs, plus three new pairs. At the final Test (4), both groups significantly improved again, and performed equally well, with three errors on average when naming the 10 warnings presented (for example, the UM group difference between Tests 3 and 4: related $t = 3.45$, $df = 11$, $p < 0.005$, one tailed test).

The identification rate of the three *new* signals was similar in both groups. In other words, the 28 subjects scored three hits/four misses on Test 3, and seven hits/three misses on Test 4. Of the four extra hits accumulated over the second session, two were usually new TBR pairs – so only five of the original set were retained. Yet *all* subjects scored, on average, 4.9 hits/2.1 misses on Test 2. Therefore there was no significant increase in the learning of the original seven alarms from Test 2 to Test 4.

Discussion

The object of the project was to review and investigate some of the factors contributing to the cognition of auditory warnings in the ICU. These alarms have an indisputable utility. However, widespread usage has also revealed a number of associated and potentially hazardous ergonomic problems, with implications both for the work of nursing and medical personnel and the safety and comfort of critically ill patients.

There may well be other, as yet unconsidered variables that affect the processing of auditory stimuli and their meanings; this study focused on three possible influences on the learning and memory of warnings: the relative distinctiveness of the sound itself, its familiarity, and the match between the perceived psychoacoustic urgency of a given alarm and the clinical significance of the event it signals (urgency mapping).

Experiment 1 measured the identification of auditory warnings by experienced ICU nurses, using stimulus sounds recorded in their own unit. Overall, the sample achieved a 63 per cent hit rate which compares very favourably with the 34 per cent and 33 per cent scored by anaesthetic clinicians in previous tests (Loeb *et al.*, 1990; Finley and Cohen, 1991). Momtahan *et al.* (1993) used two different criteria for scoring the identification performance of ICU staff in a similar study.

Their participants attained just 39 per cent when asked to name both the type of machine emitting the warning and the condition that might trigger it, but 61 per cent if only required to name either one or the other. Subjects in this project's Experiment 1 were scored on a measure akin to Momtahan *et al.*'s (1993) second, 'liberal' criterion. Nonetheless, informal post-test discussions between the Experiment 1 subjects and the researcher strongly suggested that they *could* have identified both apparatus and causative problem, if so required. Indeed, it seems unlikely that they could identify the source of an alarm without having some idea of possible triggers. The author believes that the group's relatively good performance is genuine. It should be pointed out that they were tested in an ICU room containing much of the apparatus used to provide sources of stimuli. Loeb *et al.* (1990) and Momtahan *et al.* (1993) do not clearly describe their test environments. Finley and Cohen (1991) used an area in a convention centre. It is possible, therefore, that the present sample's success is linked to the context of their test, which occurred in the (work) place where they had learnt the warnings' meanings. The machines in the test area are potentially a rich source of cues for retrieval of material. The significance of context in recall is well known (for example, Godden and Baddeley, 1975; Tulving, 1983). Tulving (1983) holds that retrieval is more likely when the information available in the test environment and the material encoded with a 'to-be-remembered' item are well matched.

Analysis of the influence of the urgency mapping variable on alarm identification in Experiment 1 showed that highly mapped pairs of warnings and meanings were almost twice as well recalled as non-UM pairs. Tulving's model indicates that appropriate cues may help direct the search for associated information in memory, and therefore that sounds of a given perceived urgency can point the way to associated events of similar perceived urgency. There are various predictable psychophysical components of perceived acoustic urgency (for example, Edworthy, 1994), and it seems that recognition of the urgency of sound occurs at an automatic, low level of cognition. Thus ICU personnel continue to exhibit startle reactions to the alarm perceived most urgent in the unit, even though they 'know' at a higher level that it indicates a relatively unimportant event. The clinician cannot but help encode auditory warnings at their various perceived urgencies. If the alarm happens to signal a situation of equivalent urgency, the encoded link of urgency is more straightforward and ecologically valid than if there is a disparity. In other words, a particular level of psychoacoustic urgency implies a meaning of similar urgency, so less logical non-UM pairs cannot be retrieved with this particular cueing pathway.

There was also a significant main effect of distinctiveness: highly distinctive warnings gave less than half the error of less distinctive alarms. This finding is predictable from research showing that the distinctiveness of an encoded trace is a key factor in memory (for example, Craik and Jacoby, 1979). Further analysis of variance revealed that there was a significant interaction between distinctiveness and urgency mapping. As might be expected, the less distinctive alarms contributed far more error than distinctive warnings when both were low UM. However, *highly distinctive*/high UM alarms produced somewhat more error than *low distinctive*/high UM warnings. The author acknowledges that one of the limitations of the project is that there is only one stimulus item per experimental cell in three-quarters of the Experiment 1 conditions. The cognition of auditory alarms in the ICU is likely to be influenced by many factors, some of which have not been accounted for in the present work.

Thus there were two unexpected simple effects: the distinctiveness/urgency mapping interaction described above; and the somewhat counterintuitive finding that error from high exposure warnings was *worse* than error from low exposure alarms, although only when these were not urgency mapped. Overall, there was no significant main effect of prior exposure. It is possible that the benefits to learning of hearing particular warnings very frequently are simply outweighed by the similarity of so many of the sounds. Nonetheless, these anomalies are difficult to explain. More research is required.

Experiment 2 examined the effect of urgency mapping on the learning of paired auditory warnings/clinical situations over two sessions. Two independent groups learnt either urgency mapped alarms/events, or non-UM pairs. After five trials, the UM subjects scored significantly more hits, but this proved to be merely a short-term advantage, as there was no difference between the groups on subsequent tests. Overall, the 28 subjects achieved 4.9 hits/2.1 misses on Test 2, but five days later they could correctly identify only three signals (with four misses). Three more trials (two with new TBR pairs) significantly improved all subjects' scores: Test 4 showed an average of seven hits/three misses. So eight trials, over two sessions totalling about an hour, taught subjects the meaning of seven auditory warnings, albeit with a sharp drop in performance after the five-day interval. It should be noted that if the subjects discarded for the purpose of comparing UM and non-UM groups are *included*, the 37 subjects identified 7.6 alarms (2.4 misses) on Test 4. This error rate is a cause for concern when considered in the context of apparatus supporting vital life functions. However, it can also be claimed that it is an encouraging result in the relatively short time involved. It illustrates the potential value of some kind of systematic training in the meaning of equipment warnings. Future work on this point could identify effective approaches to such training; for instance the

Experiment 2 participants commented favourably on the value of generating interactive imagined scenes as mnemonic aids. Nonetheless, the fluctuations in performance over time indicate that frequent active reinforcement is necessary to retain the meanings of alarms, especially if these are many in number.

Subjects did not learn any more of the original set after Test 2. Later trials were used either to relearn 'paired associates' forgotten during the interval, or to acquire new TBR pairs: two of the seven hits on Test 4 were from the three new pairs. This suggests that subjects focused on learning the novel PAs to the detriment of some of the original set, or that the new PAs displaced some of the older material from memory, even after two trials. These findings support Patterson's (1990) recommendations that auditory warnings in the workplace be restricted to a maximum of Miller's (1956) 'magical number seven'.

Concluding Remarks

This project reinforces the consensus that there needs to be a marked reduction in the number of auditory alarms in ICUs and similar areas if staff are to be able to use these devices as a positive, effective aid for patient safety and comfort. It is also clear that having a smaller, but more easily distinguished set of warning sounds will facilitate the ready identification of alarms, and enable a more rapid response to critical incidents. Kerr (1985) has suggested using a limited set of basic distinctive sounds, with each in an emergency (fast) version, and a cautionary (slow) version. Different warning sounds could be linked to particular functions for example, respiratory monitoring and ventilatory support; cardiovascular monitoring and support; renal support; 'critical' drug therapy; 'non-critical' drug therapy; fluid administration; patient comfort. This would mean, for instance, that a cardiovascular monitoring and support alarm could be refined to give either a cautionary signal (for example, ECG lead displacement), or emergency warning (for example, cessation of cardiac output). It would be important to develop such a system of seven sounds with each in two versions without there actually being 14 alarms to learn.

Previous work has suggested that urgency mapping of warnings and their meanings is a sensible ergonomic arrangement. The results of this study show that well urgency mapped alarm/meaning pairs are also better learnt (at least in the short term), and better recognised. Cognitive processing in the field is a multivariate problem. However, the author hopes that this report may be used as a starting point for further research aimed at reducing the noise bombardment in the

ICU, decreasing stress on patients and clinical staff, and making alarm systems safer and more user-friendly.

Acknowledgments

My thanks are due to Paul Barber at Birbeck College, University of London, Eva Pascoe for invaluable last-minute assistance, but especially to my wonderful wife, Irene.

References

Adam K. and Oswald, I. (1984) 'Sleep helps healing' (editorial), *British Medical Journal*, **289**(6456), 1400–01.

Baker C. (1993) 'Annoyance to ICU noise: a model of patient discomfort', *Critical Care Nursing Quarterly*, **16**(2), 83–90.

Bentley S., Murphy, F. and Dudley, H. (1977) 'Perceived noise in surgical wards and an intensive care area: an objective analysis', *British Medical Journal*, **2**(6101), 1503–06.

Bliss J., Gilson, R. and Deaton, J. (1995) 'Human probability matching behaviour in response to alarms of varying reliability', *Ergonomics*, **38**(11) 2300–12.

Cooper J. and Couvillon, L. (1983) 'Accidental breathing system disconnections', *Interim Report to the Food and Drink Administration*, Cambridge, Mass.: Arthur D Little.

Craik F. and Jacoby, L. (1979) 'Elaboration and distinctiveness in episodic memory', in L. Nilsson (ed.), *Perspectives on Memory Research: Essays in honor of Uppsala University's 500th Anniversary*, Hillsdale, NJ: Lawrence Erlbaum.

Deutsch D. (1978) 'Delayed pitch comparisons and the principle of proximity', *Perception and Psychophysics*, **23**(3), 227–30.

Edworthy J. (1994) 'Urgency-mapping in auditory warning signals', in N. Stanton (ed.), *Human Factors in Alarm Design*, London: Taylor & Francis.

Edworthy J. and Meredith, C. (1994) 'Cognitive psychology and the design of alarm sounds', *Medical Engineering and Physics*, **16**(6), 445–9.

Engen T. (1971) 'Scaling methods', in J. Kling and L. Riggs (eds), *Experimental Psychology*, London: Methuen.

Falk S. and Woods, N. (1973) 'Hospital noise levels and potential health hazards', *New England Journal of Medicine*, **289**(15), 774–81.

Finley G. and Cohen, A. (1991) 'Perceived urgency and the anaesthetist: responses to common operating room monitor alarms', *Canadian Journal of Anaesthesia*, **38**(8), 958–64.

Godden, D. and Baddeley, A. (1975) 'Context-dependent memory in two natural environments: on land and under water', *British Journal of Psychology*, **66**(3), 325–31.

Goss, A. and Nodine, C. (1965) *Paired-Associates Learning*, New York: Academic Press.

Hay, D. and Oken, D. (1972) 'The psychological stresses of intensive care nursing', *Psychosomatic Medicine*, **34**(2), 109–18.

Hellier, E., Edworthy, J. and Dennis I (1995) 'A comparison of different techniques for scaling perceived urgency', *Ergonomics*, **38**(4), 659–70.

Intensive Care Society (1983) *Standard for Intensive Care*, London: Biomedica.

Kerr, J. (1985) 'Warning devices', *British Journal of Anaesthesia*, 57, 696–708.

Kestin, I., Miller, B. and Lockhart, C. (1988) 'Auditory alarms during anesthesia monitoring', *Anesthesiology*, **69**(1), 106–9.

Loeb, R., Jones, B., Behrman, K. and Leonard, R. (1990) 'Anesthetists cannot identify audible alarms', *Anesthesiology*, **73**(3A), A539.

McIntyre, J. (1985) 'Ergonomics: anaesthetists' use of auditory alarms in the operating room', *International Journal of Clinical Monitoring and Computing*, **2**, 47–55.

Meredith, C. and Edworthy, J. (1994) 'Sources of confusion in intensive therapy unit alarms', in N. Stanton (ed.), *Human Factors in Alarm Design*, London: Taylor & Francis.

Miller, G. (1956) 'The magical number seven, plus or minus two: some limits on our capacity for processing information', *Psychological Review*, **63**(2), 81–97.

Minckley, B. (1968) 'A study of noise and its relationship to patient discomfort in the recovery room', *Nursing Research*, **17**(3), 247–50.

Momtahan, K. and Tansley, B. (1989) 'An Ergonomic Analysis of the Auditory Alarm Signals in the Operating Room and Recovery Room', Paper presented at the Annual Conference of the Canadian Acoustical Association, Halifax, Canada.

Momtahan, K., Hetu, R. and Tansley, B. (1993) 'Audibility and identification of auditory alarms in the operating room and intensive care unit', *Ergonomics*, **36**(10), 1159–76.

Noble, M. (1979) 'Communication in the ICU: therapeutic or disturbing?', *Nursing Outlook*, **27**, 195–8.

O'Carroll, T. (1986) 'Survey of alarms in an intensive therapy unit', *Anaesthesia*, **41**, 742–4.

Patterson, R. (1982) 'Guidelines for auditory warning systems on civil aircraft', CAA Paper 82017, London: Civil Aviation Authority.

Patterson, R. (1990) 'Auditory warning sounds in the work environment', *Philosophical Transactions of the Royal Society of London*, B **327**, 485–92.

Patterson, R. and Milroy, R. (1980) 'Auditory warnings on civil aircraft: the learning and retention of warnings', CAA paper 7D/S/0142 R3 London: Civil Aviation Authority.

Patterson, R., Edworthy, J., Shailer, M., Lower, M. and Wheeler, P. (1986) *Alarm Sounds for Medical Equipment in Intensive Care Areas and Operating Theatres*, Report AC598 Southampton: Institute of Sound and Vibration Research.

Redding, J., Hargest, T. and Minsky, S. (1977) 'How noisy is Intensive Care?', *Critical Care Medicine*, **5**(6), 275–6.

Smith, J. (1990) 'Psychosocial impact of the critical care environment', in C. Hudak, B. Gallo and J. Benz (eds), *Critical Care Nursing: a holistic approach*, (5th edn), Philadelphia: Lippincott.

Topf, M. and Dillon, E. (1988) 'Noise-induced stress as a predictor of burnout in critical care nurses', *Heart and Lung*, **17**(5), 567–73.

Tulving, E. (1983) *Elements of Episodic Memory*, Oxford: Oxford University Press.

Turner, A., King C. and Craddock, J. (1975) 'Measuring and reducing noise', *Hospitals*, **49**, 85–90.

Wilkins, P. (1981) 'Assessing the effectiveness of auditory warnings', *British Journal of Audiology*, **15**, 263–74.

PART VI
CONCLUSIONS

17 Key Topics in Auditory Warnings

NEVILLE A. STANTON, *University of Southampton*, and **JUDY EDWORTHY,** *University of Plymouth*

Introduction

From a practical point of view the process of auditory warning design and implementation stretches across a wide variety of domains; from a more theoretical and generic viewpoint, a range of issues are involved which tend to be fairly constant across those domains. Each chapter in this book has presented the reader with a different, yet related, mix of domain-specific problems and issues, as well as the more basic theoretical and conceptual problems which underpin these applications. In practical terms the contributions have covered the research, development and application of auditory warnings to many different domains including the following:

- air traffic control,
- aviation,
- emergency services,
- industrial plants,
- manufacturing,
- medicine,
- military,
- mining,
- motor cars and
- nuclear power.

Auditory warnings and alarms pervade both our domestic and professional lives. However examination of typical warning design and use suggests that many of the current systems do not take advantage of the body of knowledge that is already available. This book goes some way towards presenting that body of knowledge in a single volume. Other books concerned with warnings and related issues in a wider perspective are available and the reader is also referred to Lehto and Miller (1986), Stanton (1994), Laughery *et al.* (1994), and Edworthy and Adams (1996) for an overview of the whole field of warnings research.

In terms of the theoretical background to each of the chapters, dealing as they do with issues ranging from localisation of alarms to surveys of actual warning use, the focus of each chapter is inevitably at least superficially different. Yet there are similar themes throughout. In a content analysis of the contributions we have identified six central themes: problems with warnings; warning design; urgency and priority levels; auditory affordances; context dependence; and multimodality. Each of the chapters has incorporated at least one, usually several, of these themes, as can be seen from the classification by chapter shown in Table 17.1. In this the final chapter we take the opportunity of summarising the main issues centring on these themes.

Problems with Warnings

Many of the chapters, for example Chapter 8 (Burt *et al.*) suggest that auditory warnings have several advantages over their visual counterparts, and this explains why they are used in such a wide range of applications throughout industry, aviation and medicine. They reduce the need to continually monitor visual displays, they reduce visual clutter and visual workload, and they can reduce response time. As Selcon argues (Chapter 10), warnings provide a vital part of the cockpit interface, and can alert and direct the attention of the pilot to the problem when literally seconds count. It is sadly the case that this potentially vitally important medium through which information can be imparted is often not used in as effective a way as it might be.

Most contributors have made reference to the kinds of problems that tend to occur when warnings and alarms are used. Problems appear to start from the first psychoacoustic processes involved to the last action involved in responding to an alarm. Summarised below are the kinds of problems almost universally found with warnings and alarms, although it would be unlikely (and unfortunate!) if all the problems highlighted below were present in a single system. Auditory warnings and alarms are therefore typically

- set at an inappropriate level (either too loud or too quiet),
- hard to localise,
- acoustically poor and easily masked,
- irritating, insistent and startling,

- (in a set of alarms) too numerous,
- often false and unreliable,
- psychologically inappropriate,
- confusing,
- difficult to learn and remember,
- liable to divert attention away from, instead of towards, problems, and
- disruptive of verbal communications.

Table 17.1   Classification of topics by chapter

	Domains				Issues						
Authors	Emergency services	Aviation	Medical	Industrial	Problems	Urgency and priority	Context and situation	Affordances	Design	Multimodality	
Withington	●	●	●		●				●	●	Acoustics and Auditory Processes
Robinson & Casali			●		●				●		
Patterson & Datta		●					●		●		
Ballas		●			●			●	●		Auditory Cognition
Stanton & Edworthy				●	●		●	●	●		
Haas & Edworthy		●					●		●		
Burt *et al.*		●			●	●	●				
Bliss		●			●	●	●				Alarms in the Broader Context
Selcon		●					●		●	●	
Rauterberg				●	●			●		●	
Stanton & Baber				●	●				●	●	
Noyes *et al.*		●			●				●		Practical Issues
Hellier & Edworthy		●				●			●		
Meredith *et al.*			●		●						
Welch			●		●				●		

Areas

PARTS OF THE BOOK

Many of these problems can be rectified simply through better design and better thought-through implementation policies, which are dealt with in the next section. Staying with problems for a moment, at the psychoacoustic end, Withington (Chapter 2) identifies the problems concerned with pinpointing the initial source of a warning, and provides the theoretical background needed to minimise this problem. She argues that humans are capable of localising auditory warnings within five degrees of accuracy. Typically, traditional warning sounds are hard to localise because the ear cannot use either of the two mechanisms that it would normally use to localise sound because alarm frequencies tend to fall in the frequency bandwidth for which neither mechanism is reliable. Patterson's design protocol, introduced in Chapter 1, makes alarms and warnings more localisable and less resistant to masking simply by designing the basic acoustic ingredient (the pulse of sound) in such a way as to capitalise on the ears' ability to localise sound, as well as making warnings less resistant to masking by having several components in the warning sound. Localisation is clearly an important issue in auditory warning use and is becoming increasingly so in the use of 3-D displays, for example.

If the hurdles of localisation (Chapter 2) and detectability (Robinson and Casali, Chapter 3) can be overcome there are then a whole range of psychological problems needing to be dealt with, and many of the chapters focus on these. One of the most straightforward problems, but in practice one of the most intractable, is that it is often the case that there are simply too many warnings within a warning system to be reliably differentiated between by any reasonably normal listener. It is usually the case that warnings mushroom as technology develops. For example, Noyes *et al.* (Chapter 13) indicate that the current generation of modern civil passenger aircraft have approximately 450 warnings, compared to around 200 or so for the previous generation. Learning and remembering warnings and alarms thus becomes a crucial issue, and there is much evidence to show that performance in this area falls well short of optimal. The extent to which this failure in identifying warnings might be a function of the types of sounds used, rather than being due to human information-processing capacity limits, has yet to be discovered. But, however an auditory warning is designed, there is a greater risk that one warning will mask another if there are a greater number in use, unless this increase in alarm number is also accompanied by some system of prioritising warnings. Meredith *et al.* (Chapter 15) and Welch (Chapter 16) show that, even with a relatively small number of alarms within a specific work environment, people have difficulties remembering what individual warning sounds signify. The studies presented here also point up the inappropriateness and confusing nature of many of the sounds used.

Another problem seems to accompany the greater use of warnings, and this is something of a paradox. Bliss (Chapter 9) suggests that, as warning systems become more sophisticated and increase in scope and application, so the prevalence of false alarms increases. This leads to a casual attitude towards the presence of warnings which can, on occasion, be seriously mistaken. However, in the presence of an unreliable warning system, no human operator could be expected to respond to every alarm without there being consequences for other parts of the operator's activity.

All the issues of number, confusion, inappropriateness and false alarm occurrence mitigate against the use of non-verbal alarms, and of course there are alternatives such as speech alarms. In the first instance, speech-based alarms can be seen as an alternative to text-based alarms because their 'eyes-free, hands-free' aspect could make them superior to text-based warnings. However, Stanton and Baber's study (Chapter 12) shows that the transitory aspect of speech makes it prone to memory decay, and thus potentially only suited to the most urgent situations requiring immediate action. The extent to which this is also true of non-verbal alarms has yet to be explored.

Design of Warnings

Many of the chapters address the issue of designing auditory warnings and alarms. This to a large extent is the key to good human factors practice in this area. Many of the current problems associated with auditory warning and alarm use could be reduced, if not eradicated, by proper ergonomic design. It is important, however, to get the design process right from the acoustic aspects onwards, through the auditory cognitive aspects through to the more general psychological and performance aspects. In the introduction (Chapter 1) a range of guidelines was introduced. These have been embellished to a considerable extent by the contributions in this book. For example, Withington (Chapter 2) shows that, by adding broadband noise to a warning, the localisation of the sound can be considerably improved. In her study, she shows that localisation with a new, more ergonomically designed, sound was significantly better than with traditional warnings in laboratory studies. Road trials have demonstrated that this effect transfers to the real world, meaning ultimately that journey times can be reduced for emergency vehicles because their direction can be better located by other drivers, who are thus more able to take appropriate action. She is enthusiastic about transferring this finding into other domains, such as the localisation of emergency exits in public buildings and aircraft.

Patterson and Datta (Chapter 4) show how a range of acoustic techniques such as envelope filling, Nyquist whistling and fine structure doubling can be brought to bear in extending the frequency range of auditory warnings while preserving Patterson's earlier recommendations (1982) which ensure that warnings are still localisable and resistant to masking. Robinson and Casali (Chapter 3) deal with the tricky problem of being able to hear alarms while wearing ear defenders, which is often the case in practical situations. Modelling the response of the ear under such circumstances, particularly when there is some hearing loss, is crucial as a backdrop to auditory warning and alarm design too, as it will help to identify those types of design which are resistant to masking under difficult hearing conditions.

At a psychological level too there are many design issues which need to be considered. For example, some chapters recommend urgency mapped alarms, which are considered in the next section. Others, such as Rauterberg (Chapter 11), see alarms as auditory feedback and feel that the specific design of the alarms is less important than the event being signified by the alarm. This introduces the more general context of alarm and sound use and will be discussed in a later section.

One issue which often surfaces is that it can be argued that alarms are only hard to remember and inappropriate because the wrong types of sounds are being used. For example, it might be argued, why is it that we can differentiate between literally hundreds of environmental sounds whereas we seem to find it difficult to learn and remember even a dozen or so abstract alarm sounds? One approach to this problem might be to develop a methodology for eliciting information from users about their knowledge of sound, rather than having potentially circular and non-productive arguments about which types of sounds are 'best', and a methodology which might fit the bill is that presented by Stanton and Edworthy (Chapter 6). The methodology has eleven key stages:

1   establishing the need for warnings,
2   evaluating existing and modified sounds,
3   generating trial sounds,
4   appropriateness rating test,
5   design trial warning set,
6   learning/confusion test,
7   urgency mapping task,
8   design prototype warning set,
9   recognition/matching test,
10 generating standardised verbal descriptions,
11 operational test.

Stanton and Edworthy argue that, by considering the whole set of warnings together and adopting a user-centred approach, superior sets of warnings will ultimately be designed. They make no assumptions as to which types of sounds are likely to be selected, however.

Other chapters have also focused on design. For example, Hellier and Edworthy (Chapter 14) were able to validate Patterson's (1982) design methodology in their design of attensons (attention-getting sounds). Other studies show that certain design approaches can increase speed of response, which may be important in time-critical events. For example, in a study of reaction time, Haas and Edworthy (Chapter 7) identified three sound parameters that affected performance. They propose that higher pulse fundamental frequencies and higher pulse levels produce faster responses, and shorter inter-pulse intervals, along with higher fundamental frequencies and higher pulse levels, are perceived as more urgent. Here we have behavioural data which back up Patterson's earlier recommendations (1982). Other chapters show the behavioural consequences of altered (and improved) design. For example, Selcon (Chapter 10) was able to show that the design of multimodal warnings (that is, pairing visual and auditory information) led to superior performance than was achieved by either warning format presented alone. Again these results are particularly pertinent to situations that demand very fast response to the presence of a warning. Like Selcon, Stanton and Baber (Chapter 12) propose that pairing speech with visual warnings results in better performance than would be obtained by presenting speech alone, although their recommendations are for different reasons.

Several chapters highlight the need for listening tests in searching for the most appropriate warning sounds for specific applications. Patterson and Datta (Chapter 4) make the point about the importance of listening trials, and several of the other chapters describe studies which involved listening tests of one sort or another. Stanton and Edworthy (Chapter 6) have several stages of testing, listening and refinement in their methodology.

It is important too to get other psychological data from a broader context when attempting to introduce new warning systems, or to evaluate old ones. For example, Noyes *et al.* (Chapter 13) report on a study aimed at identifying problems with existing warnings as well as preferences for the next generation of warning systems. In their case study, they list the five main stages of the project: (1) familiarisation activities by the human factors team, (2) preliminary work with aircrews, (3) semi-structured interviews, (4) questionnaire survey, and (5) reality check of the findings with aircrews. Their findings suggest that most pilots would like warning systems to provide information about the secondary consequences of malfunctions as well as having some predictive capacity, and also to be able to handle multiple warnings. Bearing in mind the concerns expressed elsewhere in the book, Noyes *et al.* are cautious about developments in warning systems, and favour a human-centred perspective.

Urgency and Priority Levels

The situations which are signalled by alarms are likely to have different levels of priority; to some extent these priority levels may be fixed and to some extent they may depend on the context in which the warning is heard. In either case, it does not make much ergonomic sense to have warnings of a high acoustic urgency matched with situations of low actual importance or priority, and vice versa, simply in terms of cognitive compatibility (Edworthy, 1997). Inappropriate urgency mapping may be an important source of inappropriateness in warnings, and this has been clearly demonstrated in a number of studies (Finley and Cohen, 1991; Momtahan *et al.*, 1993; Welch, Chapter 16 of the present volume). Urgency mapping can be achieved by taking assessments of the relative priority of the situations which require alarms, rank ordering them and then designing alarms in such a way that the rank ordering of the acoustic urgency of the alarms approximately matches the rank ordering of the situations. Databases of acoustic urgency exist which will allow this kind of design and mapping to take place (for example, Edworthy *et al.*, 1991; Hellier *et al.*, 1993). A number of chapters in this book deal with the various ramifications of the concept of urgency mapping. For example, Haas and Edworthy and Hellier and Edworthy demonstrate how urgency can be designed into warnings. Haas and Edworthy (Chapter 7) show that increasing the acoustic urgency of a warning by increasing its level, pitch and speed can affect both subjective assessments of perceived urgency and the speed of response; Hellier and Edworthy (Chapter 14) show how the acoustic urgency of a set of warnings can be reliably increased and decreased without changing the nature of those alarms; and Welch (Chapter 16) demonstrates that, for a set of alarms actually in use in an intensive care unit, alarms which are urgency mapped are twice as well recalled as those which are not. Put together, this research shows that responses might be faster and recall might be better for urgency mapped alarms. Patterson and Datta (Chapter 4) also show how levels of priority can be developed in a set of warnings through acoustic manipulation.

Burt *et al.* (Chapter 8) were also able to establish a link between the level of cortical arousal of human participants and the urgency of warning sounds. However this chapter also shows that the context within which warnings are heard will have an effect on people's perception of urgency, as we shall see in a later section. Taken together, Chapters 4, 7, 8, 14 and 16 suggest that urgency is a useful concept in auditory warnings and can influence the behaviour of the person, in terms of their level of arousal and ability to identify the warning, and their speed of response to the warning. Broader contextual issues remain to be explored.

Auditory Affordances

Alarms and auditory warnings have a fairly specific role in most contexts. It is important also to look at the broader acoustic context in which warnings might be placed, and some of the chapters in the book draw our attention to this broader context. For example, Ballas (Chapter 5) presents us with a model of everyday sound perception and explores how 'everyday' listening might differ from attending specifically to warnings. One aspect of warning use which we tend to overlook is that, while listening to alarms, we are also attending to other sounds, and Ballas makes us aware that, for example, pilots in the cockpit are inevitably hearing other sounds as well as warning and alarm sounds, and that the ways in which these two classes of sound might be listened to might differ in some way. In fact, the majority of sounds will be of the non-warning, rather than the warning, sort. Meredith *et al.* (Chapter 15) provide further evidence for this, showing a 24-hour acoustic 'picture' of sounds in an intensive care unit. The vast majority of sounds heard are, inevitably, not alarm or warning sounds at all but 'normal', environmental sounds which can usually be readily identified. Ballas (Chapter 5) argues that natural sounds might be employed in warning systems, but we need to understand how people interpret them. He argues that classification systems may help drive research in this endeavour, and shows that non-speech warnings might serve five linguistic functions:

1 exclamation (for example, loud noise),
2 deixis (for example, noise from engine),
3 simile (for example, continue until popping noise stopped),
4 metaphor (for example, similar to rifle being fired),
5 onomatopoeia (for example, bang!).

This leads logically to questions about the way alarms and warnings are typically used and attended to. Rauterberg (Chapter 11) provides evidence which shows that performance in a complex task using a simulator is improved when both auditory and visual feedback are used. Auditory feedback is also very useful in providing moment-by-moment feedback on current processes and tasks. To some extent the nature of the sounds themselves does not matter; what is crucial is the association of particular sounds with particular events. Rauterberg found that the presence of sound led to better performance. It seems that participants were using sound to tell them about dynamic change in the status of the plant and to indicate when abnormal events had occurred. In this way the sound performed two functions: as a trend monitor and as an auditory warning. The presence of sound led to fewer requests for status reports, marginally better performance and greater perceived locus of control. We suspect that sound helped close the task–activity feedback loop, enabling people to track the work rate of the machines without the need for continuous visual reference.

The notion that, when we hear sounds, our natural predisposition is to try to identify the objects or events making those sounds, rather than paying conscious attention to the acoustic structure of the sound, is one well articulated by Gaver (for example, Gaver, 1993). The extent to which alarm and warning sounds are listened to in this way is relatively unexplored and depends partly on whether or not alarms and warnings are listened to in a different way from most other sounds, an issue explored by Ballas (Chapter 5). Another related issue is the idea that the use of more realistic and environmental sounds might function well as warnings. In support of this view is that it might be easy to recognise and identify many different sounds; against it is the worry that such sounds would not function sufficiently well as warning sounds, partly because of the way that they might be listened to and partly because they may not be differentiable from other, more general, background sounds. Also there is the possibility that the rapidly developing events of an alarm-generating situation may not be sufficiently rapidly signalled by everyday, environmental sounds. It is important thus to distinguish between the use of alarm-type (and other) sounds as providers of monitoring information, and the more specific use of auditory warnings and alarms as signallers of urgent information.

Whatever the nature of the distinction, it is eminently possible that the range of warnings which might function appropriately has not been fully explored as yet. It is more than likely that some types of sounds will be appropriate for some types of situations, whereas others might be more appropriate for others. For specific situations, certain types of sounds (possibly environmental, possibly abstract like a typical alarm) may have affordances for the listener (Gibson, 1979). To accompany their design protocol aimed at achieving a closer match between listener and alarm, Stanton and Edworthy (Chapter 6) propose a theory of *auditory affordances* which might be used to predict how people will act upon hearing warnings. This theory postulates a connection between the sound and its potential for action. The authors argue that certain sounds might be used to trigger a certain kind of action, thus circumventing the need for identification of the sound and its diagnosis. Within the class of iconic sounds which are likely to supply the bulk of alarms designed through the design process put forward, Stanton and Edworthy identify three basic classes of sounds: (1) nomic (for example, heartbeat for ECG monitor), (2) symbolic (for example, nursery chime for infant warmer), (3) metaphoric (for example, bubbles for syringe pump). Whether such sounds can function as alarms and warnings remains to be seen, and the development of the methodology and design of new sets of alarms could provide interesting data. Iconic sounds of this sort are quite unlike the traditional auditory warnings and it may be difficult to get them accepted by existing staff. It is also quite possible, and indeed likely, that new sets of warnings designed through the procedures proposed will end up being a mixture of traditional abstract sounds and iconic sounds. The data provided in Chapter 6 does indicate that new staff might find non-traditional warnings considerably easier to learn, however. Indeed it might be possible to extend the size of warning sets beyond the currently recommended six or seven sounds, although the distinguishability of alarm sounds, as opposed to other sounds in the environment, will need to be carefully considered if larger numbers of warning sounds are to be used.

Context Dependence

There are many issues of context which need to be looked at when considering alarm and warning use. Some of this context will come internally, from the listener's experience of the alarms themselves, their own auditory experience and abilities, and their experience of the task being undertaken. Some of the context will come externally, from the design and use of the alarm itself and the work context, which interacts with the hearer's internal context. In reality alarms and auditory warnings may be used for anything from signalling a life-threatening event to being simply a backdrop to general activity or giving feedback to the operator on line, and the chapters in this book cover this range of use. We have also used the terms 'auditory warning' and 'alarm' freely and interchangeably because there is no standardised and agreed way of differentiating between the type of sound which might be used to signal a life-threatening situation and that used as auditory feedback. Bliss (Chapter 9) presents the reader with one potential method, differentiating between 'alert', 'alarm' and 'warning', and another method of differentiation emerging across some of the chapters is the use of either 'alarm' or 'auditory warning' to signify a sound which signals an alarm condition, and the term 'trendson' to signify more general, background auditory information which does not generally serve as an alarm as such. The term 'icon' has been frequently used for this type of sound elsewhere (for example, Gaver, 1989). Although taxonomies and methods of naming exist, it is probably true to say that, as yet, there is no completely satisfactory and comprehensive method. Establishing some kind of system would seem to be an important research aid, for then we would be able to see more clearly the distinct contrast that exists between work looking specifically at high-priority alarms aimed at eliciting immediate action (such as that of Haas and Edworthy, Chapter 7) and that looking at the more general use of sound as a provider of feedback, as considered in the chapter by Rauterberg (Chapter 11). The most appropriate place to look in generating such a taxonomy would be the types of behaviour intended to be elicited by the alarm or warning.

There are many contextual issues surrounding alarm and warning design and implementation, not least of which is the fundamental issue of false alarms. Bliss considers this issue, showing that people match their response rate to alarm reliability. He also shows that there are interaction effects between false alarms, alarm criticality and task importance. Highly critical warnings were responded to even if the false alarm rate was quite high, independently of task importance. This demonstrates that people are able to adapt quite quickly to the environmental context in which they find themselves. Stanton and Edworthy (Chapter 6) also suggest that the context may have an important influence upon what people expect the warning sound to be, particularly the environment in which the sounds are learnt.

Another contextual issue is that of the actual urgency of an alarm in comparison to its acoustic urgency. The study by Burt *et al.* (Chapter 8) illustrates this point. The authors found that, although participants were able to differentiate between three alarms differing acoustically in their urgency prior to performance on a task, this differentiability was clouded if the warnings were not matched appropriately to the situations they were signalling. Thus, if an acoustically highly urgent alarm was used to signal a situation known to have only low situational urgency, the effect of the urgency in the alarm itself was lost. This was also shown in the responses to the situations themselves, where performance was determined by participants' knowledge of task priority, and not by the design of the alarm. This demonstrates the importance of context in alarm response and accentuates the relationship between perceived, acoustic urgency and the urgency ascribed to the situation itself, often referred to as 'situational urgency'. Whilst there are benefits to learning and retention, as well as cognitive compatibility, to urgency mapping, as discussed earlier in this chapter, the relationship between cues inherent in the warning (such as urgency) and cues inherent in the situation being signalled (such as its known, or predicted, urgency) needs to be further explored in research. Either way, Burt *et al.*'s study shows the importance of situational knowledge in judging the importance of events. Rauterberg proposes that the sounds used do not matter in themselves because it is the events that they are signalling which need to be interpreted, and then acted upon. From this perspective it would be interesting to see if the manipulation of the sounds has any effect in this kind of experimental scenario. For example, non-urgent and urgent alarms could be matched to urgent and non-urgent situations within a process control scenario in a 2 x 2 design; alarms could be swapped from one situation to another, and so on. What is clear is that the impact of specific designs on performance needs to be further explored.

To a degree, almost all of the chapters deal with some of the issues of context in alarm and auditory warning use in some way or another. Warnings are always context-dependent, and it may often be the case that what is important on one occasion may be trivial on another. Ultimately the whole working context needs to be considered, but here we move from alarm- and warning-specific research and implementation issues to those more appropriately placed within the general mainstream of ergonomics and human factors.

Multimodality

Some of the chapters (Chapters 2, 10, 11 and 12) deal to some degree or another with the issue of multimodality. This is where more than one sensory mode is used to provide warning information, usually simultaneously. There are a number of interesting issues underlying this design concept, as well as some practical and specifically warnings-related issues. Normally the senses act together to interpret the world, and there is a considerable amount of theoretical work which directly addresses the issue of multisensory stimulation. On the topic of localisation, Withington (Chapter 2) suggests that, when stimulation is provided in more than one mode, the effect is often greater than the sum of the parts. One of the principal reasons for this is that we make assumptions about the unity of objects when stimulation from more than one sense appears to coalesce (for example, Welch and Warren, 1980) and, unless information appears to contradict this conclusion, this is the one that we will make. Thus, if we see and hear information at the same time, we will assume that the different elements are coming from the same source if they show similar characteristics. This 'assumption of unity' (Welch and Warren, 1980) is clearly applicable in the case of warning localisation, as we can use information in an auditory signal to draw attention to a visual event. One area where this has a major application is in three-dimensional auditory displays. As well as the specific issue of localisation, the use of auditory icons to draw attention to visual events was also explored by Rauterberg (Chapter 11). Here the performer of the task may be able to make assumptions about the specific location of problems through the use of appropriate iconic sounds, through direct mapping at a cognitive level (although not necessarily at a perceptual, localisation level) between environmental, 'real' sounds and the images and affordances associated with these sounds. The potential use for such sounds has been explored throughout the book, and there are implications too for such sounds from a cross-modal or multimodal perspective.

From a slightly different perspective, the issue of using more than one mode for warning presentation was considered in Chapter 12 (Stanton and Baber), who considered the relative merits of speech- and text-based warnings. Whilst their results are somewhat ambivalent, there is other research in the warnings area which tends to support the view that warnings presented auditorily will produce higher levels of compliance than those presented visually (Wogalter and Young, 1991; Wogalter *et al.*, 1993). These same studies also show that, when warnings are presented in both modes, compliance levels are higher still. A similar finding was shown by Selcon (Chapter 10), where performance was improved through stimulating more than one sense simultaneously.

While studies tend to show that increased stimulation, particularly across modes, tends to lead to improved performance, it is essential to consider the effect that such stimulation will have within the whole context of the tasks for which warnings are an accompaniment. For example, if stimulating a person in two modes increases interference on the primary or secondary tasks that he or she is performing, it might be argued that such stimulation is excessive except in the most high-priority situations; using this kind of level of stimulation for relatively low-urgency situations and conditions might be inappropriate from the viewpoint of cognitive compatibility. There is not necessarily a cost to be paid, however; a study by Sorkin *et al.* (1988) showed that more sophisticated and multimodal warning systems did not necessarily have a deleterious effect on workload and performance when the whole pattern of performance on the experimental task was considered. This may not be true for all applications, and the key here may be to use such methods sparingly, and to think through (and evaluate empirically) the effects of such a warning strategy. Certainly using more than one mode to warn appears attractive both in terms of the performance levels that can be obtained, and in view of the fact that in our everyday lives our senses work together in understanding the world.

References

Edworthy, J. (1997) 'Cognitive compatibility and warning design', *International Journal of Cognitive Ergonomics*, **1**(3), 193–209.

Edworthy, J. and Adams, A.S. (1996) *Warning Design: A Research Prospective*, London: Taylor & Francis.

Edworthy, J., Loxley, S.L. and Dennis, I.D. (1991) 'Improving auditory warning design: relationship between warning sound parameters and perceived urgency', *Human Factors*, **33**(2), 205–31.

Finley, G.A. and Cohen, A.J. (1991) 'Perceived urgency and the anaesthetist: responses to common operating room monitor alarms', *Canadian Journal of Anaesthesia*, **38**(8), 958–64.

Gaver, W.W. (1989) 'The SonicFinder: an interface that uses auditory icons', *Human–Computer Interaction*, **4**, 67–94.

Gaver, W.W. (1993) 'How in the world do we hear? Explorations in ecological acoustics', *Ecological Psychology*, **5**, 283–313.

Gibson, J.J. (1979) *The Ecological Approach to Visual Perception*, New York: Houghton Mifflin.

Hellier, E.J., Edworthy, J. and Dennis, I.D. (1993) 'Improving auditory warning design: quantifying and predicting the effects of different warning parameters on perceived urgency', *Human Factors*, **35**(4), 693–706.

Laughery, K.R., Wogalter, M.S. and Young, S.L. (1994) *Human Factors Perspectives on Warnings*, Santa Monica, CA: Human Factors and Ergonomics Society.

Lehto, M.R. and Miller, J.M. (1986) *Warnings Volume 1: Fundamentals, Design and Evaluation Methodologies*, Ann Arbor: Fuller Technical Publications.

Momtahan, K.L., Hetu, R. and Tansley, B.W. (1993) 'Audibility and identification of auditory alarms in operating rooms and an intensive care unit', *Ergonomics*, **36**, 1159–76.

Patterson, R.D. (1982) 'Guidelines for auditory warning systems on civil aircraft', CAA paper 82017, London: Civil Aviation Authority.

Sorkin, R.D., Kantowitz, B.H. and Kantowitz, S.C. (1988) 'Likelihood alarm displays', *Human Factors*, **30**, 445–9.

Stanton, N.A. (1994) *Human Factors in Alarm Design*, London: Taylor & Francis.

Welch, R.B. and Warren, D.H. (1980) 'Immediate perceptual response to intersensory discrepancy', *Psychological Bulletin*, **88**(3), 638–67.

Wogalter, M.S. and Young, S.L. (1991) 'Behavioural compliance to voice and print warnings', *Ergonomics*, **34**, 78–89.

Wogalter, M.S., Kalsher, M.J. and Racicot, B.M. (1993) 'Behavioural compliance with warnings: effect of voice, context and location', *Safety Science*, **16**(5/6), 637–54.

Subject Index

Author Index